Workbook for
Structure and Function of the Human Body

Workbook for Structure and Function of the Human Body

Fourth Edition

Ruth Lundeen Memmler, M.D.
*Professor Emeritus, Life Sciences;
formerly Coordinator, Health, Life Sciences, and Nursing,
East Los Angeles College, Los Angeles, California*

Dena Lin Wood, R.N., B.S., P.H.N.
*Staff Nurse, Memorial Hospital of Glendale,
Glendale, California*

Illustrated by Anthony Ravielli

J.B. Lippincott Company   Philadelphia
London   Mexico City   New York   St. Louis   São Paulo   Sydney

Sponsoring Editor: Patricia Cleary
Developmental Editor: Joyce Mkitarian
Design Director: Tracy Baldwin
Cover Photo Illustration: Eric Pervuknin and Kathy Ziegler
Cover Designer: Anthony Frizano
Production Superviosr: J. Corey Gray
Production Editor: Rosanne Hallowell
Compositor: Circle Graphics
Text Printer/Binder: Kingsport Press
Cover Printer: The Lehigh Press, Inc.

Copyright © 1987, by J.B. Lippincott Company.
Copyright © 1983, 1977, 1972 by J.B. Lippincott Company.

All rights reserved. No part of this book may be used or reproduced in any manner whatsoever without written permission except for brief quotations embodied in critical articles and reviews. Printed in the United States of America. For information write J.B. Lippincott Company, East Washington Square, Philadelphia, Pennsylvania 19105.

ISBN 0-397-54608-4

6 5 4 3

Preface

Workbook for Structure and Function of the Human Body, Fourth Edition, is designed to assist the beginning student in learning the basics of human anatomy and physiology. While it will be most effective when used in conjunction with the fourth edition of *Structure and Function of the Human Body,* it is also applicable to other basic textbooks on the subject.

The continued emphasis on physiology is reflected in the exercises and problems presented in this revision. There is maximum coordination with the parent textbook, and the chapter sequence follows that of the textbook. The Practical Applications portions of the workbook use clinical situations to test the student's understanding of a subject. While focusing on disease conditions, they require answers based on knowledge of normal structure and function.

The exercises were prepared with the aim of helping the student to learn, and not merely to test knowledge. A certain amount of repetition has been purposely incorporated as a means of reinforcement.

Contents

1. Introduction to the Human Body *1*
2. Chemistry, Matter, and Life *11*
3. Cells and Their Functions *19*
4. Tissues, Glands, and Membranes *27*
5. The Skin *37*
6. The Skeleton—Bones and Joints *43*
7. The Muscular System *61*
8. The Nervous System *71*
9. The Sensory System *91*
10. The Endocrine System and Hormones *103*
11. The Blood *109*
12. The Heart *119*
13. Blood Vessels and Blood Circulation *127*
14. The Lymphatic System and Immunity *145*
15. Respiration *155*
16. Digestion *163*
17. Metabolism, Nutrition, and Body Temperature *179*
18. The Urinary System and Body Fluids *185*
19. Reproduction *199*
20. Heredity *215*
21. Biological Terminology *219*

Workbook for
Structure and Function of the Human Body

1

Introduction to the Human Body

I. Overview

Living things are organized from simple to complex levels. The simplest living form is the **cell**, the basic unit of life. Specialized cells are grouped into **tissues** that in turn are combined to form **organs**; these organs form **systems**.

The systems include the skeletal system, the framework of the body; the muscular system, which moves the bones; the circulatory system, consisting of the heart and blood vessels that transport vital substances; the digestive system, which converts raw food materials into products usable by cells; the respiratory system, which adds oxygen to the blood and removes carbon dioxide; the integumentary system, the body's covering; the urinary system, which removes wastes and excess water; the nervous system, the central control system which includes the organs of special sense; the endocrine system, which produces the regulatory hormones; and the reproductive system, by which new individuals of the species are produced.

All of the cellular reactions that sustain life together make up **metabolism,** which can be divided into **catabolism** and **anabolism**. In catabolism, food is broken down into smaller molecules with the release of energy. This energy is stored in the compound **ATP** (adenosine triphosphate) for use by the cells. In anabolism, simple compounds are built into substances needed by the cells.

All the systems work together to maintain a state of balance or **homeostasis**. The main mechanism for maintaining homeostasis is negative feedback, by which the state of the body acts to keep conditions within set limits.

It is essential that a special set of terms be learned in order to locate parts and to relate the various parts to each other. Imaginary lines called **planes of division** separate parts of the body into regions in much the same way that the Equator, the Tropics of Cancer and Capricorn, and the Arctic and Antarctic Circles divide the earth into zones. Further divisions of the earth by lines of latitude and longitude make it possible to pinpoint locations accurately. Similarly, separation into areas and regions within the body, together with the use of the special terminology for directions and locations, makes it possible to describe an area within the human body with great accuracy.

The large internal spaces of the body are the **cavities**, in which various organs are located. The **dorsal cavities** are further divided into the cranial cavity and the spinal canal. The **ventral cavities** are the thoracic and abdominal cavities. The lower portion of the abdominal cavity is the pelvic cavity.

The metric system is used for all scientific measurements. If you learn to "think metric" you will find it easier to use than the older systems used in the United States because it is based on multiples of 10.

II. Topics for Review

A. Body systems
B. Body processes
C. Body directions
D. Body cavities
   a. Dorsal and ventral cavities
   b. Regions in abdominal cavity
E. The metric system

III. Matching Exercises

Matching only within each group, print the answers in the spaces provided.

Group A

extracellular   anabolism   physiology
organs   homeostasis   cells
systems   tissues

1. The building phase of metabolism is   _____

2. A combination of specialized groups of cells forms   _____

3. The study of how the body functions is   _____

4. A combination of various tissues forms parts having a special
   function that are called   _____

5. Several different parts and organs grouped together for specific
   functions form   _____

6. The basic units of life are called   _____

7. A state of balance in the body is   _____

8. Fluids located outside the cells are described as   _____

Group B

epigastric region   ventral   distal
umbilicus   lateral   medial
proximal   thoracic region   transverse

1. To indicate nearness to the midsagittal plane, use the word   _____

2. A part that is away from the midline (or toward the side) is _____

3. To indicate that a part is near or toward the point of origin, use _____

4. A part that is away from the point of origin is _____

5. A horizontal or cross section is also said to be _____

6. The central region of the abdomen just below the breast bone is the _____

7. Another name for the navel is the _____

8. The upper or chest portion of the ventral body cavities is the _____

9. The word that means *toward the belly surface* is _____

Group C

caudal   cranial   posterior
sagittal plane proximal   transverse plane

1. To say *toward the origin of a part,* use the word _____

2. To indicate that a part is toward the rear use _____

3. The word that means nearer the tail region is _____

4. To indicate that a part is nearer the head use the word _____

5. A plane that divides the body into superior and inferior parts is a(n) _____

6. A plane that divides the body into left and right parts is a(n) _____

Group D

urinary system   integumentary system   skeletal system
endocrine system   reproductive system   respiratory system

1. The system that includes the hair, nails, and skin is the _____

2. The bones, joints, and related parts form the system called the _____

3. Another name for the excretory system is the _____

4. The system of scattered organs that produce hormones is called the _____

5. The system that includes the sex organs is the _____

6. The lungs and bronchial tubes form the system called the _____

Group E

spinal canal   cranial cavity   diaphragm
negative feedback   nervous system   pelvic cavity

1. The cavity in the lower part of the abdominal cavity is the   _____

2. The muscular partition between the two ventral body cavities is the   _____

3. A system that controls and coordinates the body is the   _____

4. The upper part of the dorsal body cavity is the   _____

5. The lower part of the dorsal body cavity is the   _____

6. A mechanism for maintaining homeostasis is   _____

IV. Labeling

For each of the following illustrations, print the name or names of each labeled part on the numbered lines.

1. _____
2. _____
3. _____
4. _____
5. _____
6. _____
7. _____

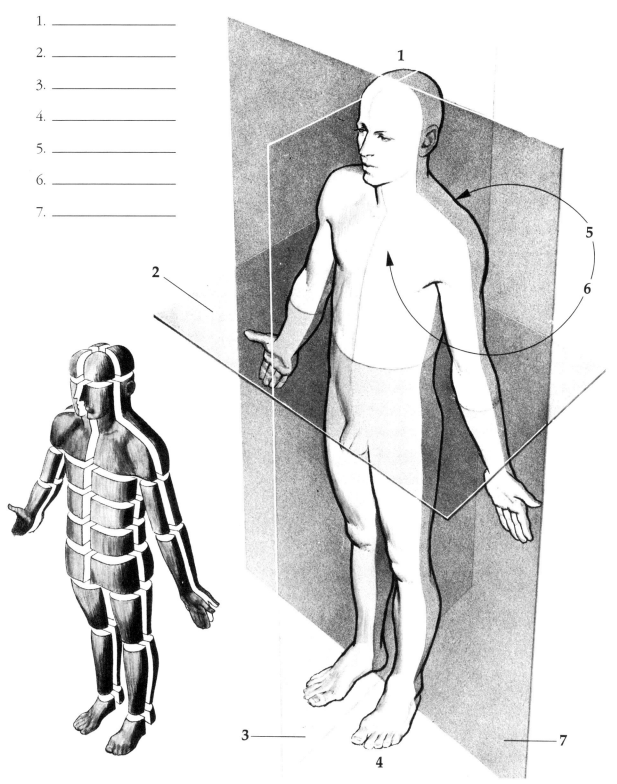

Body planes and directions

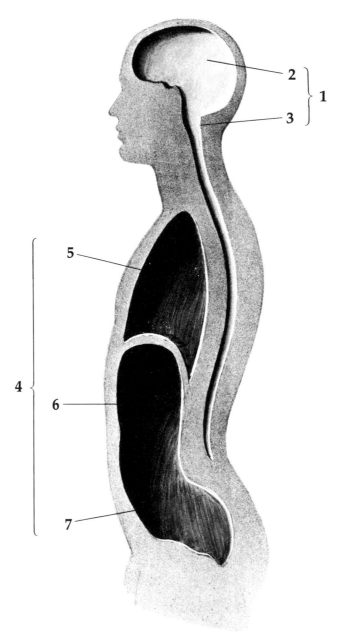

Side view of body cavities

1. _____
2. _____
3. _____
4. _____
5. _____
6. _____
7. _____

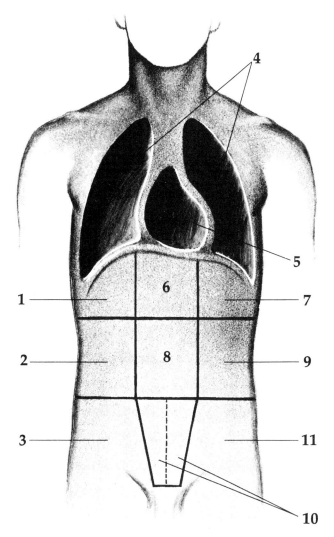

Front view of body cavities and the regions of the abdomen

1. _____
2. _____
3. _____
4. _____
5. _____
6. _____
7. _____
8. _____
9. _____
10. _____
11. _____

V. Completion Exercise

Group A

Print the word or phrase that correctly completes each sentence.

1. The energy compound of the cell is _____

2. Groups of specialized cells are organized into _____

3. Regions and directions in the body are described according to the position in which the body is upright, with the palms facing forward. This is called the _____

4. The midline plane that divides the body into right and left halves is the _____

5. Planes that divide the body into upper and lower parts are called _____

6. The plane that divides the body into front and rear parts is the _____

7. The space that encloses the brain and spinal cord forms one continuous cavity, the _____

8. The space that houses the brain is the _____

9. The elongated canal that contains the spinal cord is known as the _____

10. The ventral body cavities include an upper space containing the lungs, the heart, and the large blood vessels, which is called the _____

11. The lower ventral body cavity is quite large and is called the _____

12. The large ventral body cavities are separated from each other by a muscular partition, the _____

13. The large lower ventral body cavity may be subdivided into nine regions, including three along the midline. The uppermost of these midline areas is the _____

14. All of the chemical reactions that maintain the cell make up _____

Group B

Print the word that correctly completes each sentence about the metric system.

1. The standard unit for measurement of volume, slightly greater than a quart, is a(n) _____

2. The number of grams in a kilogram is _____

3. The number of centimeters in an inch is _____

4. The number of milliliters in 0.5 liters is _____

5. The standard unit for measurement of length is the _____

VI. Practical Applications

Study each discussion. Then print the appropriate word or phrase in the space provided.

Group A

1. The gallbladder is located just below the liver. The directional terms that best describe this relationship include _____

2. The kidneys are located behind the other abdominal organs. This relationship may be described as _____

3. The tips of the fingers and toes are farthest from the region of origin of these digits, so they are said to be the most _____

4. The entrance to the stomach is nearest the point of origin or beginning of the stomach, so this part is said to be _____

5. The ears are located away from the midsagittal plane or toward the side, so they are described as being _____

6. The head of the pancreas is nearer the midsagittal plane than its tail portion, so the head part is more _____

7. The diaphragm is above the abdominal organs; it may be described as _____

Group B

On the ward in which postoperative patients are being cared for you are asked to study certain cases and answer the following questions.

1. Mr. A had an appendectomy. The area of the abdomen in which the appendix is located is in the lower right side and is known as the _____

2. Mrs. B had a history of gallstones. The operation to remove these stones involved the upper right part of the abdominal cavity, or the _____

3. Miss C was injured in an automobile accident. In addition to a number of fractures, she suffered a ruptured urinary bladder. The area involved, in the lower midline part of the abdomen, was the _____

4. Mr. B required an extensive exploratory operation that necessitated incision through the navel. This portion of the abdomen is the _____

2

Chemistry, Matter, and Life

I. Overview

Chemistry is the physical science that deals with the composition of matter. To appreciate the importance of chemistry in the field of health, it is necessary to know about atoms, molecules, elements, compounds, and mixtures. Though exceedingly small particles, atoms possess a definite structure: the **nucleus** contains **protons** and **neutrons**, and surrounding the nucleus are the **electrons**. An **element** is a substance consisting of just one type of atom. Union of two or more atoms produces a **molecule**; the atoms may be alike (such as the oxygen molecule) or different (sodium chloride, for example), and in the latter case the substance is called a **compound**. To go a step further, a combination of compounds, each of which retains its separate properties, is a **mixture** (salt water is one example). Chemical compounds are constantly being formed, altered, broken down, and recombined into other substances. Water is a vital substance composed of hydrogen and oxygen. It makes up more than half of the body and is needed as a solvent and a transport medium. Hydrogen, oxygen, carbon, and nitrogen are the elements that constitute about 99% of living matter, while calcium, sodium, potassium, phosphorus, sulfur, chlorine, and magnesium account for most of the remaining 1%. Proteins, carbohydrates, and lipids are among the compounds formed from these elements.

II. Topics for Review

A. Atoms and elements
B. Molecules and compounds
C. Water, solutions, and suspensions
D. Chemical bonds
E. Acids and bases
F. Organic compounds

III. Matching Exercises

Matching only within each group, print the answers in the spaces provided.

Group A

chemicals	organic	nucleus
chemistry	atoms	pharmacology

1. The smallest complete units of matter are called _____

2. The science that deals with the composition of all matter is _____

3. The study of all aspects of drugs is called _____

4. Aspirin, penicillin, and all other drugs are classified as _____

5. The part of the atom containing most of its mass including protons and neutrons is the _____

6. The chemical compounds that characterize living things are described as _____

Group B

molecule	electrons	mixture
neutrons	elements	protons
compounds		

1. The positively charged particles inside the atomic nucleus are _____

2. The noncharged particles within the atomic nucleus are _____

3. The negatively charged electric particles outside the atomic nucleus are the _____

4. Substances composed of one type of atom are called _____

5. The unit formed by the union of two or more atoms is the _____

6. Substances that result from the union of two or more different atoms are known as _____

7. The combination of various compounds that remain intact and retain their properties is designated a(n) _____

Group C

carbohydrates	cations	acid
electrolytes	anions	water
proteins	buffer	pH

1. Positively charged ions are called _____

2. Negatively charged ions are _____

3. Compounds that form ions when in solution are called _____

4. Compounds of nitrogen, carbon, oxygen, and hydrogen are called _____

5. Simple sugars are classified as _____

6. A substance which helps to maintain a stable hydrogen ion concentration in a solution is a(n) _____

7. The universal solvent is _____

8. The symbol for hydrogen ion concentration is _____

9. A substance that donates a hydrogen ion to another substance is a(n) _____

Group D

element   covalent   phospholipids
suspension   colloidal   amino acid
atomic number   solution

1. Nitrogen is an example of a(n) _____

2. An element can be identified by its _____

3. A building block of proteins is a(n) _____

4. The group of lipids that contains phosphorus in addition to carbon, hydrogen, and oxygen is the _____

5. A chemical bond formed by the sharing of electrons is called _____

6. A mixture in which substances will settle out unless the mixture is shaken is a(n) _____

7. Cytoplasm and blood plasma are examples of a type of suspension described as _____

8. Salt water is an example of a(n) _____

Group E

ionic   decomposed   neutral
carbon   lipid   enzyme

1. A pH of 7.0 is termed _____

2. A type of protein that acts as a catalyst in metabolic reactions is a(n) _____

3. Another name for a fat is a(n) _____

4. An electrolyte is formed by a bond described as _____

5. Elements cannot be changed into something else by physical or chemical methods; that is, they cannot be _____

6. Organic chemistry is based on the element _____

IV. Labeling

For each of the following illustrations, print the name or names of each labeled part on the numbered lines.

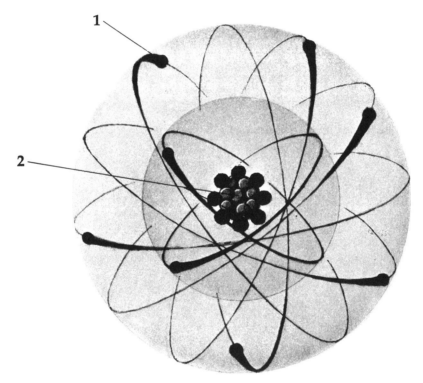

Oxygen atom

1. _____   2. _____

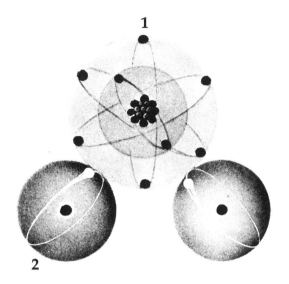

Molecule of water

1. _____   2. _____

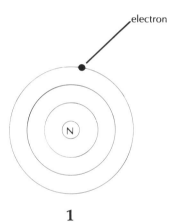

1

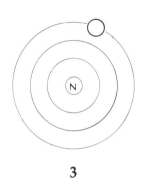

3

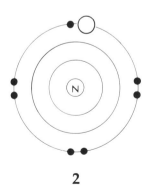

2

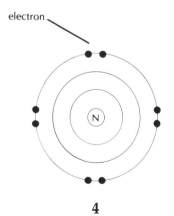

4

Cations and anions

1. _____   3. _____

2. _____   4. _____

V. Completion Exercise

Print the word or phrase that correctly completes each sentence.

1. The four elements that make up about 99% of living cells are carbon, hydrogen, and oxygen plus _____

2. Most people keep a shaker of salt on the table. Salt is an example of a combination of two different elements. Such a combination is called a(n) _____

3. Compounds first found in living organisms, as for example starch in potatoes, are classified as _____

4. An element that is part of the air we breathe also is part of the water we drink. This element is _____

5. The smallest particle of salt obtainable that would still have the properties of salt is the _____

6. The salt in salt water will regain its properties if the water is boiled. Since water and salt do not combine chemically, this solution is an example of a(n) _____

7. If we could remove a single electron from a sodium atom, or could add a single electron to a sodium atom, the result would be an atom with either a positive or a negative charge, known as a(n) _____

8. Numerous essential body activities are possible owing to the property of certain compounds to form ions when in solution. Such compounds are called _____

9. The name given to a chemical system that prevents changes in hydrogen ion concentration is _____

VI. Practical Applications

The following medical tests are based on the principles of chemistry and physics.

1. Mr. B complained of shortness of breath. Several studies were done including a visible tracing of the electric currents produced by his heart muscle. Such a record is called a(n) _____

2. Joan, age 4, was brought to the clinic by her mother because she experienced attacks of fainting and unconsciousness. As an aid in diagnosis, a graphic record of her brain's electric current was obtained. This brain wave record is called a(n) _____

3. A routine test done on Ms. J showed glucose in her urine—an abnormal finding. Glucose is one of a group of compounds found in certain foods and classified as _____

4. Analysis of Mr. K's urine showed the presence of albumin. Albumin is an example of compounds found in the body that contain nitrogen, carbon, hydrogen, and oxygen. These compounds are classified as _____

3

Cells and Their Functions

I. Overview

The cell is the basic unit of life; all life activities result from the activities of cells. The study of cells began with the invention of the light microscope and has continued with the development of electron microscopes. Cell functions are carried out by specialized structures within the cell called **organelles**. These include the nucleus, ribosomes, mitochondria, Golgi apparatus, and endoplasmic reticulum (ER).

An important cell function is the manufacture of **proteins**, including enzymes (organic catalysts). Protein manufacture is carried out by the ribosomes in the cytoplasm according to information coded in the deoxyribonucleic acid (DNA) of the nucleus. DNA is also involved in the process of cell division or **mitosis**. Before cell division can occur, the DNA must double itself so that each daughter cell produced by mitosis will have exactly the same kind of DNA as the parent cell.

The cell membrane is important in regulating what enters and leaves the cell. Some substances can pass through the membrane by **diffusion**, which is simply the movement of molecules from an area where they are in higher concentration to an area where they are in lower concentration. Water can diffuse rapidly through the membrane by the process termed **osmosis**. For this reason, cells must be kept in solutions that have the same concentrations as the cell fluid. If the cell is placed in a solution of higher concentration (a hypertonic solution) it will shrink; in a solution of lower concentration (a hypotonic solution) it will swell and may burst. The cell membrane can also selectively move substances into or out of the cell by **active transport**, a process that requires energy (ATP) and carriers. Large particles and droplets of fluid are taken in by the processes of **phagocytosis** and **pinocytosis**.

II. Topics for Review

A. Cell structure
B. Protein synthesis
C. Cell division (mitosis)
D. Movement of materials across the cell membrane

III. Matching Exercises

Matching only within each group, print the answers in the spaces provided.

Group A

semipermeable	isotonic	diffusion
enzymes	mitosis	active transport
osmosis	filtration	

1. The protein substances that assist in the cell's chemical reactions are the _____

2. The process of body cell division is known as _____

3. The membrane of the cell is said to be _____

4. The spread of molecules throughout an area is known as _____

5. Water molecules diffuse through the cell membrane by the process of _____

6. The passage of solutions through a membrane as a result of mechanical force is called _____

7. A solution that has the same concentration of molecules as the fluids within the cell is described as _____

8. The cell uses energy to move substances across the membrane in a process called _____

Group B

mitochondria	ribosomes	cilia
lysosomes	nucleolus	cell membrane
flagellum	ER	

1. A system of membranes throughout the cell is the _____

2. A small globule within the nucleus is the _____

3. Small bodies in the cytoplasm that act in the manufacture of proteins are _____

4. The outer covering of the cell is the _____

5. The organelles that convert energy to ATP are _____

6. A long whiplike projection used in cell locomotion is a(n) _____

7. Small bodies in the cell that contain digestive enzymes are _____

8. Small hairlike projections from the cell used to create movement around the cell are called _____

Group C

ATP nucleotide centriole
DNA genes RNA

1. The chemical in the nucleus that makes up the chromosomes is _____

2. The organelle that is active in cell division is the _____

3. The energy compound of the cell is _____

4. A building block of nucleic acids is a(n) _____

5. The hereditary factors in the cell are _____

6. The nucleic acid that carries information from the nucleus to the ribosomes is _____

Group D

edema pinocytosis osmotic pressure
hypertonic hypotonic active transport

1. A cell may take in so much water it will burst if it is placed in a solution that is _____

2. A cell takes in droplets by the process of _____

3. The force that draws water into a solution is called _____

4. A solution with a salt concentration greater than 0.9% is described as _____

5. An accumulation of fluid in the tissues is called _____

6. The cell membrane is described as selectively permeable because it can carry out _____

IV. Labeling

For each of the following illustrations, print the name or names of each labeled part on the numbered lines.

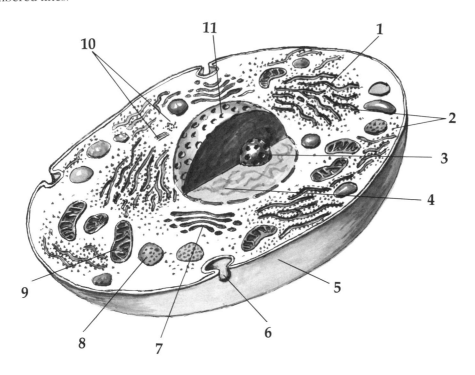

A typical cell

1. _____
2. _____
3. _____
4. _____
5. _____
6. _____
7. _____
8. _____
9. _____
10. _____
11. _____

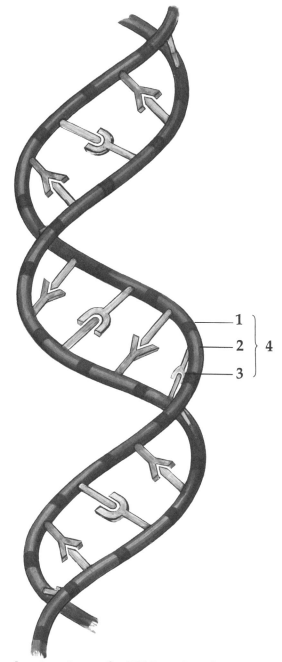

The basic structure of a DNA molecule

1. _____

2. _____

3. _____

4. _____

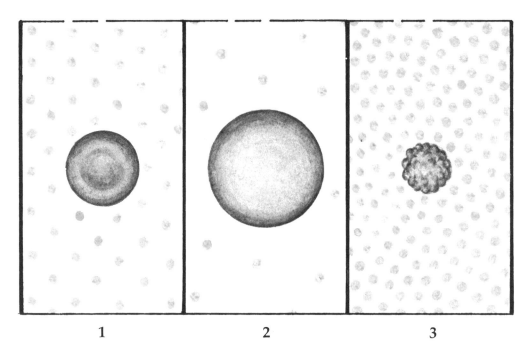

1	2	3

Osmosis. Describe the solutions shown in each picture

1. _____

2. _____

3. _____

V. Completion Exercise

Print the word or phrase that correctly completes each sentence.

1. The general term that describes all the chemical reactions by which food is made usable by the cells is _____

2. Groups of similar cells that function together for the same general purpose make up _____

3. The metric unit that is used to measure cells is the _____

4. Small structures within a cell that perform special functions are called _____

5. The control center of the cell which contains the chromosomes is the _____

6. Chromosomes are composed mainly of _____

7. The energy compound of the cell is _____

8. Before a cell can divide the chromosomes must _____

9. The number of daughter cells formed when a cell undergoes mitosis is _____

VI. Practical Applications

Study each discussion; then print the appropriate word or phrase in the space provided.

Observations you might make while touring a hospital laboratory include the following:

1. The janitor in the laboratory was using a cleaning solution that contained ammonia. You will recall that this would cause ammonia molecules to spread throughout the room. This movement of molecules from an area of high concentration to other areas where concentration is low is called _____

2. One of the laboratory technicians was trying to separate solid particles from a liquid mixture. He poured the mixture into a paper-lined funnel. The liquid flowed through the funnel while the solids remained behind on the paper. This process is called _____

3. A laboratory worker was carefully measuring certain salts in order to prepare a normal saline solution. Normal saline is used to replace lost body fluids because the concentration is nearly the same as that inside the cells. Such a solution is said to be _____

4. While doing a complete blood count a technician noted that some of the red blood cells had ruptured. The solutions used were tested to determine whether they were too dilute. Osmosis of water into a cell could be the cause of cell breakage. When a red blood cell bursts it is said to _____

5. A student was learning how to do blood smears. Upon examination of the blood with the microscope he found that many red blood cells appeared shrunken. The explanation was that he was proceeding so slowly that the liquid part of the blood was evaporating, leaving a highly concentrated solution. Such a solution is described as being _____

4

Tissues, Glands, and Membranes

I. Overview

The cell is the basic unit of life. Individual cells are grouped according to function into **tissues**. The four main groups of tissues include **epithelial tissue**, which forms glands, covers surfaces, and lines cavities; **connective tissue**, which gives structure and holds all parts of the body in place; **nerve tissue**, which conducts nerve impulses; and **muscle tissue**, which produces movement.

The simplest combination of tissues is a **membrane**. Membranes serve several purposes, a few of which are mentioned here: they may serve as dividing partitions; may line hollow organs and cavities; and may anchor various organs. Membranes that have epithelial cells on the surface are referred to as epithelial membranes. Two types of epithelial membranes are mucous membranes, which line passageways leading to the outside, and serous membranes, which line body cavities and fold over the internal organs.

II. Topics for Review

A. Classification of tissue
B. Functions of the four main groups of tissues
C. Types of glands
D. Epithelial membranes

III. Matching Exercises

Matching only within each group, print the answers in the spaces provided.

Group A

cartilage	tissues	bone
adipose	cilia	squamous
exocrine	layered	secretions

1. Groups of cells similar in structure and function are called _____

2. An important function of epithelium is the production of _____

3. Dust and other foreign particles are moved along the airways by tiny hairlike projections from epithelium called _____

4. The type of connective tissue that stores fat and serves as a heat insulator is called _____

5. One of the hard connective tissues that has the important function of acting as a shock absorber and as a bearing surface to reduce friction between moving parts is _____

6. Osseous tissue is similar to cartilage in its cellular structure. In development, cartilage gradually becomes impregnated with calcium salts to form _____

7. Flat, irregular epithelial cells are described as _____

8. The term *stratified* means _____

9. Glands that secrete through ducts are called _____

Group B

ligaments	marrow	neuron
mucus	collagen	transitional
cartilage	secretions	fascia

1. Mucus, digestive juices, and sweat are examples of _____

2. Dust and other inhaled foreign particles are trapped in a secretion called _____

3. Layers of fibrous connective tissue around muscles are called _____

4. A crepelike type of epithelium that is capable of great expansion is called _____

5. The strong connective bands that support joints are called _____

6. The main fibers in connective tissue are made of a flexible white protein called _____

7. The tough, elastic substance found at the ends of long bones is _____

8. Blood cells are produced in the red _____

9. Another name for a nerve cell is a(n) _____

Group C

myocardium   neurilemma   myelin
voluntary muscle   connective tissue   visceral muscle
fibers

1. Areolar, adipose, and osseous tissue all act as the body's supporting fabric and are therefore classified as _____

2. The basic structural unit of nerve tissue, the neuron, consists of a nerve cell body plus small branches, which are called _____

3. The ability of certain nerves to repair themselves is due to the presence of a thin coating membrane called the _____

4. Some nerve fibers, like telephone wires, are encased in a protective covering, or sheath. This fatty insulating material is called _____

5. The thickest layer of the heart wall is formed by cardiac muscle or _____

6. Muscle tissue is classified into three types. That which forms the walls of the organs within the ventral body cavities is called _____

7. Skeletal muscle is usually under the control of the will. It is therefore described as _____

Group D

suture   connective tissue   periosteum
epithelium   tendon

1. The tissue that forms a protective covering for the body and that lines the intestinal tract and the respiratory and urinary passages is called _____

2. Repair of damaged nerve and muscle tissue is accomplished by the growth of _____

3. A layer of fibrous connective tissue around a bone is the _____

4. The size of a scar following the healing of a clean wound may best be reduced by bringing the edges together with a(n) _____

5. A band of connective tissue that connects a muscle to a bone is a(n) _____

Group E

pleura   membrane   mucous membranes
pericardium   lubricants   fascia
peritoneum   serous membranes

1. Any thin sheet of material that separates two or more groups of substances is classified as a(n) _____

2. The membranes that line the closed cavities within the body are _____

3. The tough membranes composed entirely of connective tissue which serve to anchor and support organs are the _____

4. The linings of tubes and spaces that are connected with the outside are largely epithelial. They are _____

5. The membrane that covers each lung is known as the _____

6. The special sac that encloses the heart is known as the _____

7. The serous membrane of the abdominal cavity is the largest of its kind and is called the _____

8. An important function of most epithelial membranes is to produce fluids that serve as _____

Group F

superficial fascia	periosteum	synovial membranes
parietal layer	perichondrium	mucous membranes
mesothelium	capsules	cutaneous membrane

1. Membranous connective tissues that enclose organs are called _____

2. The tough connective tissue membrane that serves as bone covering is the _____

3. Covering cartilage is a membrane similar to that covering bone. It is called _____

4. Secretions produced by the lining of joint cavities act as lubricants to reduce friction between the ends of bones. These linings are _____

5. The linings of the various parts of the respiratory tract are all _____

6. The tissue that underlies the skin is known as the _____

7. The part of a serous membrane that is attached to the wall of a cavity or sac is the _____

8. The type of epithelium that covers serous membranes is called _____

9. The skin is described as the _____

IV. Labeling

Print the name or names of each labeled part on the numbered lines.

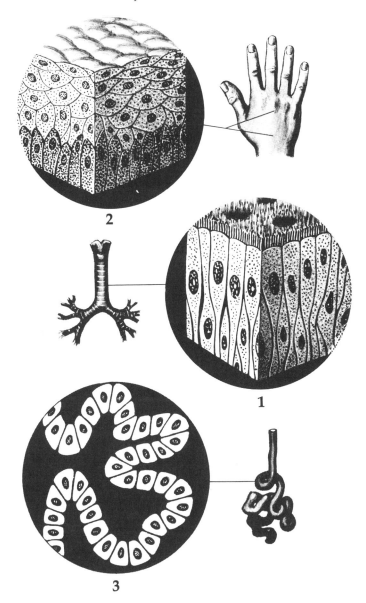

Three types of epithelium

1. _____

2. _____

3. _____

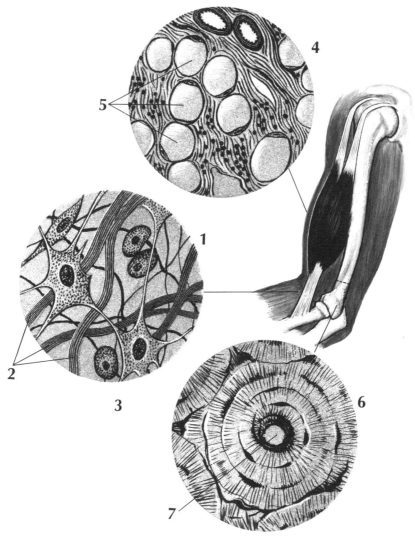

Connective tissue

1. _____   5. _____

2. _____   6. _____

3. _____   7. _____

4. _____

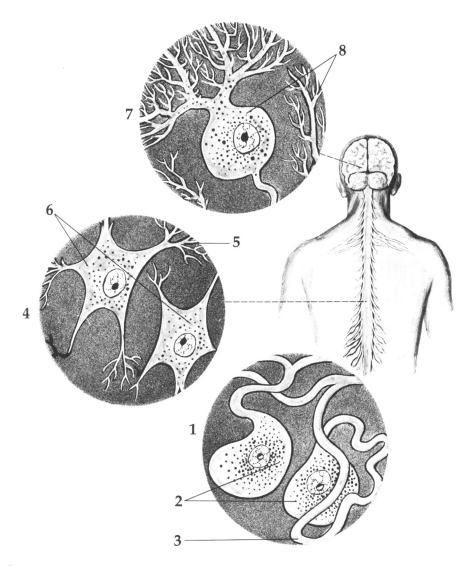

Nerve tissue

1. _____ 5. _____

2. _____ 6. _____

3. _____ 7. _____

4. _____ 8. _____

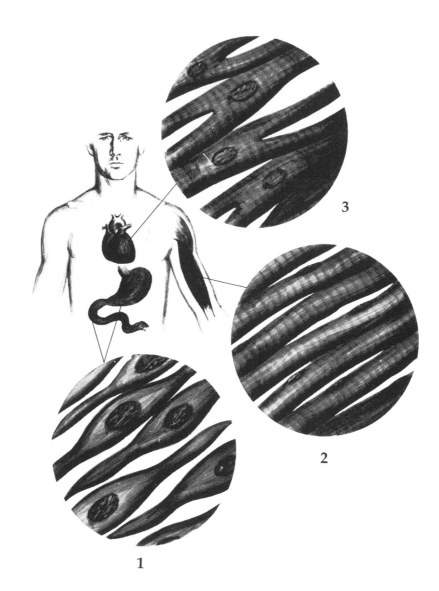

Muscle tissue

1. _____

2. _____

3. _____

34

V. Completion Exercise

Print the word or phrase that correctly completes each sentence.

1. Another term for the smooth involuntary muscle of most hollow organs is _____

2. The supporting tissue of the body organs is called _____

3. The basic unit of nerve tissue is the nerve cell, the scientific name for which is _____

4. Movement is produced by the tissue known as _____

5. The noun that indicates a layer of serous membrane is _____

6. The lubricant produced by membranes that line cavities connected with the outside is known as _____

7. The microscopic hairlike projections found in the cells lining most of the respiratory tract are called _____

8. The layer of a serous membrane that lines the wall of a cavity or sac is called the _____

9. Layers of fibrous connective tissue that enclose certain internal organs are called _____

10. The tough connective tissue membrane that covers most parts of all bones is given the name _____

11. A lubricant that reduces friction between the ends of bones is produced by the _____

VI. Practical Applications

Study each discussion. Then print the appropriate word or phrase in the space provided.

Group A

The following patients were seen in the hospital emergency room.

1. Little J, age 7, fell while riding his bicycle. He sustained several gashes on his face. At the emergency center the cuts were cleansed and sutured. Suturing is desirable because it reduces the size of the connective tissue replacement, called the _____

2. Student K, age 16, had sustained a deep cut on his index finger while he was whittling a piece of wood. Although the cut was sutured and healed well, he noted loss of sensation on the tip of the cut finger. However, after a few weeks the sensation began to return. This was possible because of the thin nerve fiber covering that aids in the nerve repair process. This membrane is called _____

Group B

While observing in an outpatient clinic a student noted the following cases.

1. Baby J experienced difficulty in breathing and had a copious discharge from his nose. A diagnosis of URI (upper respiratory infection) was made. The location of the membrane and the type of discharge indicated that the involved membrane was one of the _____

2. Mrs. K complained of a swelling in the left groin. She had suffered previously from an infection of bone in the middle back; now it appeared that the infection had traveled along the fibrous covering of some of the back muscles. Such muscle coverings are called _____

3. Mr. B was concerned about swelling and tenderness over his neck and upper back. His work involved the demolition of old buildings; he had become careless about personal cleanliness. Infection now involved the skin and connective tissue under it. The "sheet" that underlies the skin is called _____

4. Mrs. J had suffered a painful bump on her ankle. The swelling involved the superficial tissues and the fibrous covering of the bone, or the _____

5. Mrs. C had undergone extensive surgery because of deformities due to rheumatoid arthritis, an inflammatory disorder of the membranes lining the joint spaces. These lining membranes are known as _____

6. Ms. G experienced abdominal pains following longstanding infection of the pelvic organs. Connective tissue bands (adhesions) were found to extend throughout the peritoneal surface. The layer of peritoneum that is attached to the organs is called the _____

7. Student N suffered a mild concussion while playing football and it was feared that there might be damage to the brain coverings. These brain and spinal cord coverings are known as _____

8. Mrs. J was quite ill. Her symptoms were those associated with the disease called lupus erythematosus. She complained that it hurt to breathe because the membranes covering the lungs were involved. These membranes are called the _____

5

The Skin

I. Overview

Because of its various properties, the skin can be classified as an **enveloping membrane**, an **organ**, and a **system**. A cross section of skin reveals its layers of **epidermis** (the outermost layer), **dermis** (the true skin where the skin glands are mainly located), and the **subcutaneous tissue** (the underlayer).

The skin serves the essential functions of **protecting** deeper tissues against drying and against invasion by harmful organisms, **regulating** body temperature, and **obtaining information** from the environment. It also **excretes** water, salts, and some waste in the form of sweat. The pigment **melanin** gives the skin its color; races that have been exposed to the tropical sun for thousands of years have highly pigmented skin. The protein **keratin** in the epidermis thickens and protects the skin, and **sebum**, secreted by the sebaceous glands, lubricates the skin and prevents dehydration. Hair and nails, composed mainly of keratin, are structures associated with the skin.

Being the most visible aspect of the body, the skin is the object of much quackery, and vast sums of money are spent in efforts to beautify it; good general health is, however, the most important part of skin health and beauty.

II. Topics for Review

A. Skin layers
   1. Epidermis
   2. Dermis
   3. Subcutaneous layer
B. Skin glands
   1. Sudoriferous glands
   2. Sebaceous glands
C. Hair and nails
D. Functions of the skin

III. Matching Exercises

Matching only within each group, print the answers in the spaces provided.

Group A

epidermis	sebaceous glands	melanin
keratin	sudoriferous glands	dermis
integument	subcutaneous layer	connective tissue

1. In its role as a system, the skin is called the _____

2. The protein in the epidermis that thickens and protects the skin is _____

3. Certain glands produce sweat; these are the _____

4. The tissue layer under the skin is the _____

5. The oily secretion on skin and hair is produced by _____

6. Several layers of epithelial cells form the outermost part of the skin, the _____

7. Since the epidermis is lacking in blood vesels, nutritive substances reach the epidermal cells from the underlying _____

8. The framework of the dermis is composed of _____

9. Skin color is due largely to the presence of the pigment called _____

Group B

dilate	absorption	ciliary glands
pathogens	infection	fat
receptors	dermis (or corium)	

1. In its function of regulating body temperature the skin dissipates heat as the blood vessels enlarge or _____

2. Modified sweat glands are found in the eyelid edges. These are known as _____

3. In its role of protecting deeper tissues the skin prevents drying and invasion by _____

4. The skin's function of obtaining information from the environment is due to the presence of a variety of sensory nerve endings. One general term for these is _____

5. The subcutaneous tissue is composed of connective tissue and _____

6. The nerve endings of the skin are located mainly in the _____

7. Medications are given by mouth or by injection more often than they are applied to the skin. This is because the skin has limited powers of _____

8. Following a wound or injury of the skin, pathogens may enter and cause a(n) _____

Group C

    intact   follicle   sebum
    melanin   nerve endings   temperature
    corneum   ceruminous   germinativum

1. Dilation of blood vessels brings more blood to the surface so that heat is dissipated into the air. This is one way in which the skin acts to regulate _____

2. The skin is an able defender against invasion by pathogens as long as it remains unbroken or _____

3. Exposure to sunlight causes an increase in the quantity of the pigment _____

4. The uppermost layer of the epidermis is the stratum _____

5. The modified sweat glands that secrete wax are called _____

6. The oily secretion of the sebaceous glands is _____

7. Obtaining information about the environment is a function of the skin's _____

8. The deepest layer of the epidermis, which contains living, dividing cells, is the stratum _____

9. Each hair develops within a sheath called a(n) _____

IV. Labeling

For the following illustration, print the name or names of each labeled part on the numbered lines.

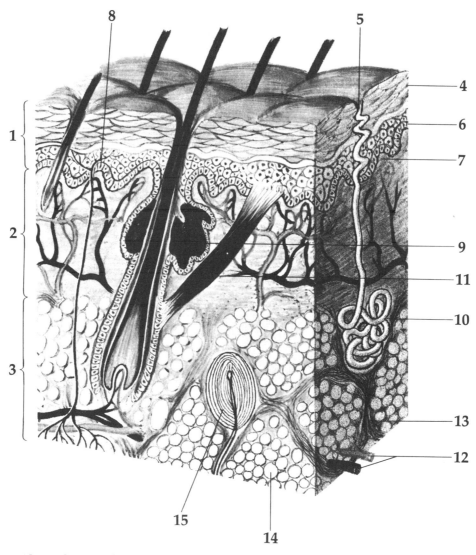

Cross section of the skin

1. _____
2. _____
3. _____
4. _____
5. _____
6. _____
7. _____
8. _____
9. _____
10. _____
11. _____
12. _____
13. _____
14. _____
15. _____

V. Completion Exercise

Print the word or phrase that correctly completes each sentence.

1. The outer cells of the epidermis, which are constantly being shed, are designated the horny layer, or _____

2. The pigment of the skin is _____

3. One means by which the body protects itself against pathogens is production of skin secretions which are slightly _____

4. The nails originate from the outer part of the _____

5. The ceruminous glands and the ciliary glands are modified _____

6. Hair and nails are composed mainly of the protein that thickens and protects the skin, called _____

7. The blood vessels which nourish the epidermis are located in the skin layer just below the epidermis, called the _____

VI. Practical Applications

Study each discussion. Then print the appropriate word or phrase in the space provided.

1. L, age 15, consulted a skin doctor with his father. The son's skin was marked by pimples and blackheads and had a roughened appearance. This common disorder, called acne vulgaris, is found mainly in adolescents. It involves infection of the oil producing glands of the skin called the _____

2. Mr. M, a laborer, had neglected to give his skin proper care. He now had numerous painful boils in his axillae (armpits). These boils were caused by the entrance of bacteria into the sheaths in which the hairs grow. These sheaths are called _____

3. There was also a deep-seated infection of the tissue under the skin on Mr. M's lower back. This tissue lying under the skin is described by the term _____

4. Mt. Laurel hospital used footprints to identify all newborn babies in the nursery. The pattern of a footprint is formed by elevations and depressions in the skin layer beneath the epidermis, the layer called the _____

5. Ms. J, age 17, consulted her family physician because she had noticed scattered dark areas on the surface of her skin. The doctor thought her problem was due simply to too much sun exposure. The skin pigment that gives the skin color and increases when one is exposed to sunlight is _____

6

The Skeleton—Bones and Joints

I. Overview

The skeletal system protects and supports the body parts and serves as an attachment for the muscles, which furnish the power for movement. The skeletal system includes some 206 bones; the number varies slightly according to age and the individual.

Bones are composed of living tissue and have their own systems of blood and lymphatic vessels and nerves. Bone tissue may be either **spongy** or **compact**. Compact bone is found in the shaft (diaphysis) of a long bone and in the outer layer of other bones. Spongy bone makes up the ends (epiphyses) of a long bone and the center of other bones. **Red marrow** occurs in certain parts of all bones and manufactures the blood cells; **yellow marrow**, which is largely fat, is found mainly in the central cavities of the long bones.

Bone is produced by cells called **osteoblasts** which gradually convert cartilage to bone during development. The mature cells that maintain bone are called **osteocytes**, and the cells that break down bone for remodeling and repair are the **osteoclasts**.

The entire bony framework of the body is called the skeleton. It is divided into two main groups of bones, the **axial skeleton** and the **appendicular skeleton**. The axial skeleton includes the skull, spinal column, ribs, and sternum. The appendicular skeleton consists of the bones of the arms and legs, the shoulder girdle, and the pelvic girdle.

A **joint** is the region of union of two or more bones; joints are classified on the basis of the degree of movement permitted. Connective tissue bands, the **ligaments**, hold the bones together in all the freely movable (synovial) joints and many of the less movable joints.

II. Topics for Review

A. Structure of bone
B. Bone cells
C. Bone formation
D. Functions of bones
E. Bones of the axial skeleton

F. Bones of the appendicular skeleton
G. Landmarks on bones
H. Types of joints

III. Matching Exercises

Matching only within each group, print the answers in the spaces provided.

Group A

    cartilage   red marrow   yellow marrow
    periosteum   calcium salts   appendicular skeleton
    axial skeleton   endosteum   osteoblast

1. The bony framework of the head and trunk forms the   _____

2. Production of blood cells is carried on mainly in the   _____

3. The combination of bones that form the framework for the extremities is called the   _____

4. The fatty material found inside the central cavities of long bones is   _____

5. The tough connective tissue membrane that covers bones is   _____

6. The somewhat thinner membrane that lines the central cavity of long bones is   _____

7. The pliability of the young child's bones is due to their relatively large proportion of   _____

8. A cell that produces bone is called a(n)   _____

9. The brittleness of the old person's bones is due to their relatively large proportion of   _____

Group B

    cranium   occipital bone   parietal bones
    ethmoid bone   sutures   sphenoid bone
    temporal bones

1. The delicate spongy bone located between the eyes is called the   _____

2. The bone which forms the back of the skull, and part of the base of the skull, is the   _____

3. That part of the skull which encloses the brain is the   _____

4. The bat-shaped bone that extends behind the eyes and also forms part of the base of the skull is the   _____

5. The paired bones that form the larger part of the upper and side walls of the cranium are the _____

6. The two bones that form the lower sides and part of the base of the central areas of the skull are _____

7. The cranial bones join at places called _____

Group C

| mandible | maxillae | zygomatic bone |
| hyoid | nasal bones | lacrimal bone |

1. At the inside corner of each eye is a very small bone, the _____

2. The only movable bone of the skull is the _____

3. Lying just below the skull proper is a U-shaped bone called the _____

4. The higher part of each cheek is formed by a bone called the _____

5. The two bones of the upper jaw are the _____

6. The two slender bones that form much of the bridge of the nose are the _____

Group D

true ribs	cervical region	floating ribs
lumbar region	diaphysis	thoracic region
rib cage	coccyx	scoliosis

1. The shaft of a long bone is the _____

2. The spinal column is divided into five regions; the first seven vertebrae comprise the main framework of the neck, the _____

3. The third section of the vertebral column consists of five bones that are somewhat larger than the first 19 vertebrae; these form the _____

4. The second part of the vertebral column has 12 bones which make up the _____

5. In the child the tail part of the vertebral column is made of four or five small bones that later fuse; this is the _____

6. Protecting the heart and other organs as well as supporting the chest are functions of the surrounding framework called the _____

7. The first seven pairs of ribs are called the _____

8. Among the false ribs are two pairs, the last two, which are very short and do not extend to the front of the body. These are the _____

Group E

    patella   epiphysis   radius
    ulna   tibia   sesamoid
    fibula   olecranon process

1. The upper part of the ulna forming the point of the elbow is the _____

2. The medial forearm bone is the _____

3. The kneecap is also called the _____

4. Of the two bones of the leg the larger is the _____

5. The forearm bone on the thumbside is the _____

6. The patella is the largest of a type of bone that is encased in connective tissue. It is described as _____

7. The lateral bone of the leg is the _____

8. The end of a long bone is the _____

Group F

    articulation   foramina   acetabulum
    foramen magnum   anterior fontanel   sacrum
    spongy   diaphysis   diarthroses

1. The skull of the infant, being in its formative stage, has a number of soft spots. The largest of these is the _____

2. The type of bone tissue found at the ends of a long bone is described as _____

3. The region of the spinal column below the lumbar region is made of four to five fused bones. It is called the _____

4. The largest opening in the skull, containing the spinal cord, is the _____

5. Openings or holes that extend into or through bones are called _____

6. The deep socket in the hip bone that holds the head of the femur is the _____

7. The shaft of a long bone is the _____

8. The region of union of two or more bones is called a joint or _____

9. The more freely movable joints are _____

Group G

costal shoulder girdle phalanges
greater trochanter calcaneus carpal bones
metacarpal bones symphysis pubis ilium
pelvic girdle processes ligaments

1. The five bones in the palm of each hand are the _____

2. The largest of the tarsal bones is the heel bone or _____

3. The 14 small bones that form the framework of the fingers on each hand are the _____

4. In the pelvic girdle, the os coxae is divided into three areas. The upper wing-shaped part is the _____

5. The bones of the wrist are the _____

6. The clavicle and the scapula are contained in the _____

7. The os coxae articulating with the sacrum comprise the _____

8. An adjective that refers to the ribs is _____

9. The pubic parts of the two ossa coxae unite to form the joint called the _____

10. The connective tissue bands that hold bones together at joints are called _____

11. The large, rounded projection at the upper and lateral portion of the femur is the _____

12. The prominences on bones that serve as muscle attachments have the general name of _____

Group H

articular cartilage flexion rotation
synovial membrane abduction extension
ball-and-socket hinge

1. The contacting surfaces of each joint are covered by a layer called the _____

2. A bending motion that decreases the angle between two parts is _____

3. The lubricating fluid inside a joint cavity is produced by the lining of the cavity termed the _____

4. Movement away from the midline of the body is known as _____

5. Motion around a central axis is called _____

6. The reverse of flexion is _____

7. The type of joint that allows for circumduction is a(n) _____

8. The type of joint found at the elbow is a(n) _____

IV. Labeling

For each of the following illustrations, print the name or names of each labeled part on the numbered lines.

1. _____
2. _____
3. _____
4. _____
5. _____
6. _____
7. _____
8. _____
9. _____
10. _____
11. _____
12. _____
13. _____
14. _____
15. _____
16. _____
17. _____
18. _____
19. _____
20. _____
21. _____
22. _____
23. _____
24. _____
25. _____
26. _____
27. _____

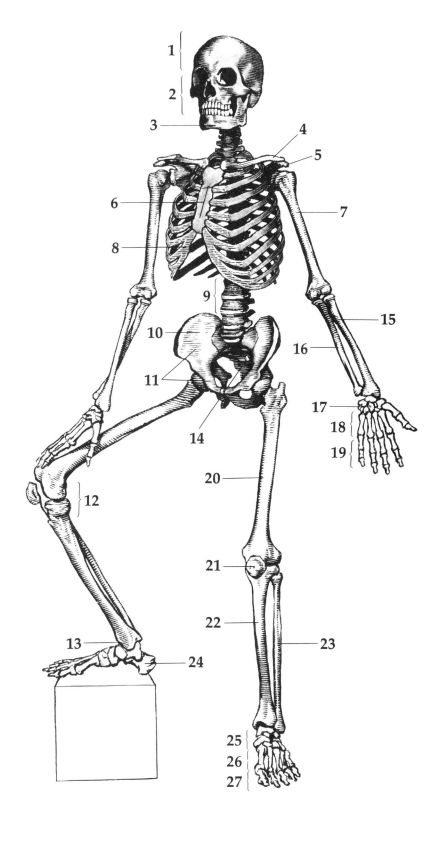

The skeleton

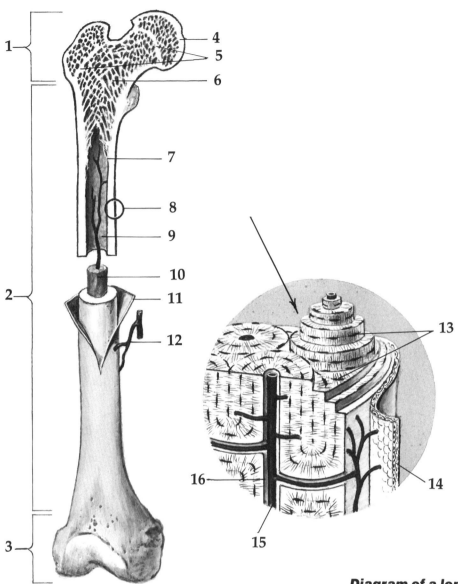

Diagram of a long bone

1. _____
2. _____
3. _____
4. _____
5. _____
6. _____
7. _____
8. _____
9. _____
10. _____
11. _____
12. _____
13. _____
14. _____
15. _____
16. _____

1. _____
2. _____
3. _____
4. _____
5. _____
6. _____

7. _____
8. _____
9. _____
10. _____
11. _____
12. _____
13. _____
14. _____
15. _____

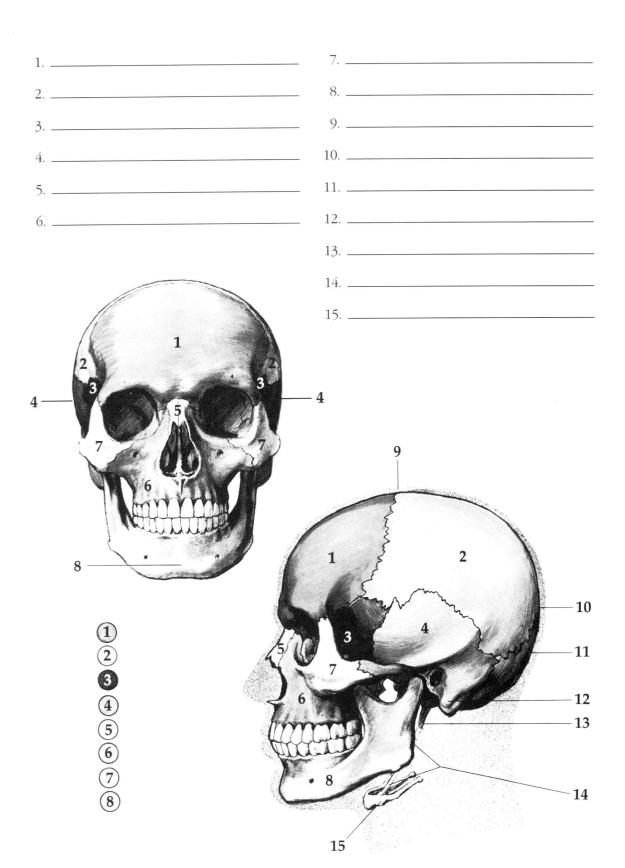

Skull from the front and from the left

51

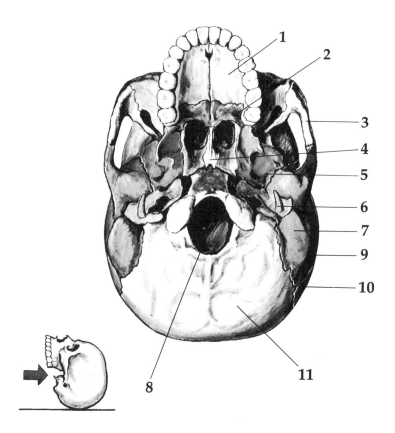

1. _____
2. _____
3. _____
4. _____
5. _____
6. _____
7. _____
8. _____
9. _____
10. _____
11. _____

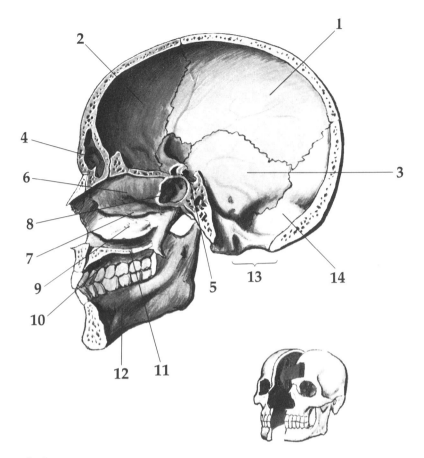

Skull, internal view

1. _____ 8. _____

2. _____ 9. _____

3. _____ 10. _____

4. _____ 11. _____

5. _____ 12. _____

6. _____ 13. _____

7. _____ 14. _____

53

1. _____
2. _____
3. _____
4. _____
5. _____
6. _____
7. _____
8. _____
9. _____
10. _____
11. _____
12. _____
13. _____

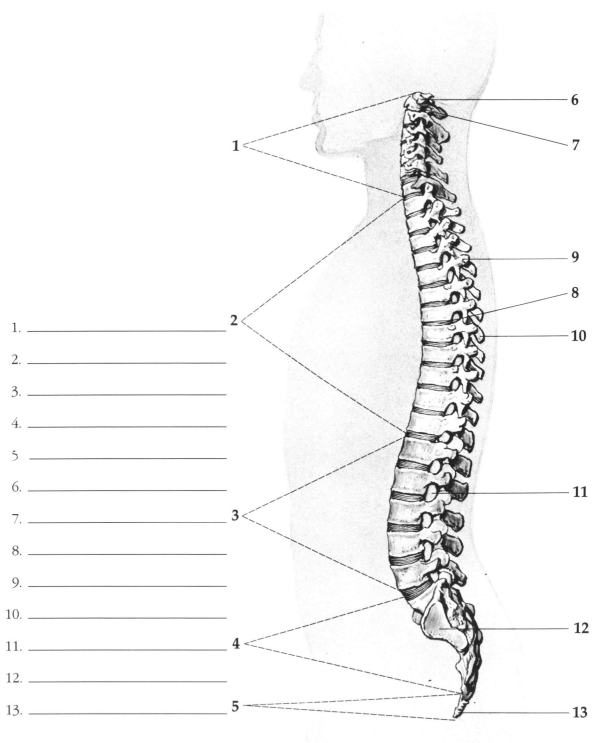

Vertebral column

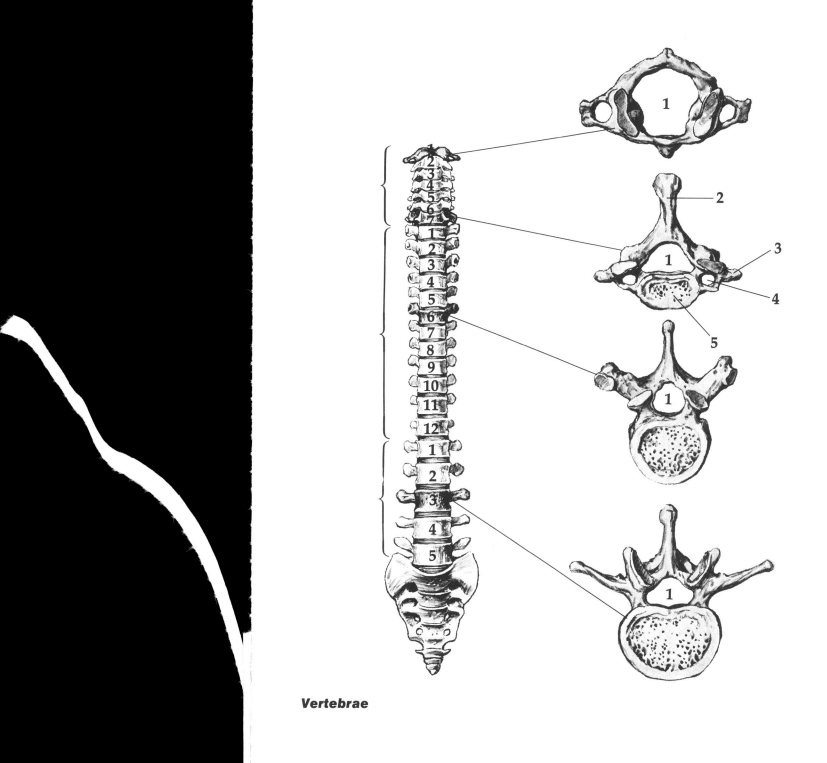

Vertebrae

1. _____ 4. _____

2. _____ 5. _____

3. _____

55

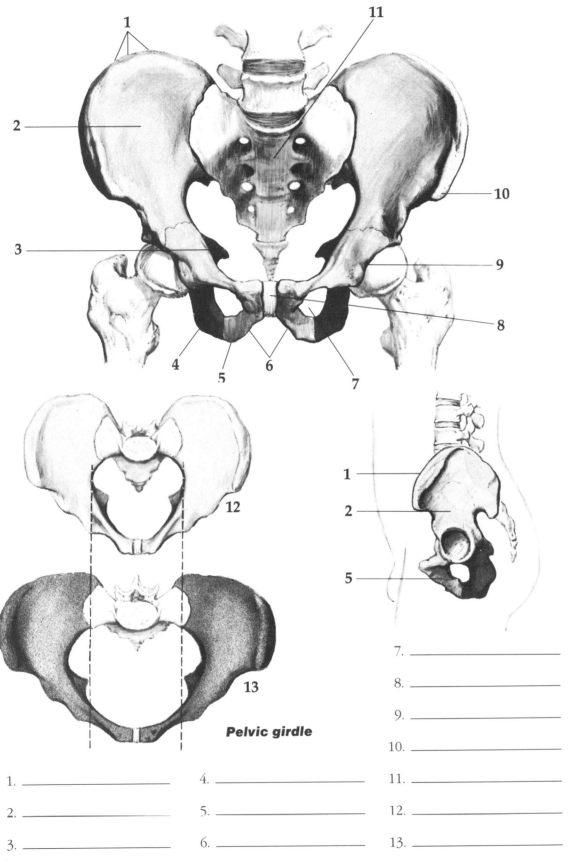

Pelvic girdle

1. _____
2. _____
3. _____
4. _____
5. _____
6. _____
7. _____
8. _____
9. _____
10. _____
11. _____
12. _____
13. _____

56

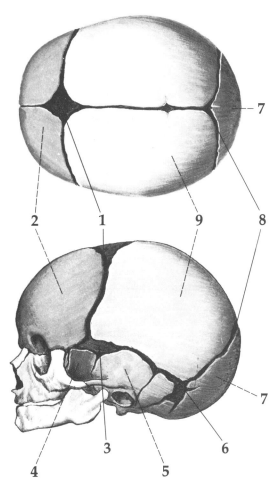

Infant skull, showing fontanels

1. _____ 6. _____

2. _____ 7. _____

3. _____ 8. _____

4. _____ 9. _____

5. _____

V. Completion Exercise

Print the word or phrase that correctly completes each sentence.

1. During development, secondary bone-forming centers appear across the ends of the long bones. Each end of a long bone is called a(n) _____

2. When bone-forming cells mature and become enclosed in hardened bone material, they are referred to as _____

3. In the embryonic stage of bone development, most of the developing bones are made of _____

4. When bone is resorbed, cells that break down bone become active; these cells are called _____

5. The type of bone tissue that makes up the shaft of a long bone is called _____

6. The hardening of a long bone begins at the center of the _____

7. The skull, vertebrae, ribs, and sternum make up the part of the skeleton called the _____

8. The cervical and lumbar curves, which appear after birth, are referred to as _____

9. A suture is an example of an immovable joint also called a(n) _____

10. Pivot, hinge, and gliding joints are examples of freely movable joints also called _____

VI. Practical Applications

Study each discussion, then print the appropriate word or phrase in the space provided.

Group A

A group of high school seniors was involved in a serious traffic accident on the way home from the prom.

1. There was a pronounced swelling of the upper right side of Mary's head. X-ray films showed a fracture of the largest skull bone, the _____

2. Mary also suffered an injury to one of the two large bones of the pelvic girdle. This bone articulates with the sacrum and is named the _____

3. John suffered multiple injuries to his left lower extremity. Protruding through the skin was a splintered portion of the longest bone in the body, the _____

4. Susan thought her injuries were the least serious, so she walked several blocks to find help. Then she noticed that her right knee was not functioning normally. Examination revealed a fractured kneecap. Another name for the kneecap is _____

5. Harry, the driver of the car, was forcibly thrown against the steering wheel. He suffered fractures of the sixth and seventh ribs, which are included among the _____

Group B

Mr. B, age 58, was admitted to the general hospital because of acute pain and swelling of his right great toe. He also complained of a chronic backache. Mr. B underwent a complete physical examination.

1. X-ray films showed involvement of the toe joints. The framework of the toes is made up of bones called the _____

2. Spurs of bony material were found to be present at the edges of the vertebrae just above the sacrum, which is the part of the spinal column called the _____

Group C

Mrs. C, age 36, visited her doctor's office because of swelling and pain in the joints of her hands and fingers. Examination revealed the following:

1. Evidence of inflammation and overgrowth of the lining membrane of the joint cavities, a membrane that is called the _____

2. Difficulty in moving the joints of the fingers due to damage to the normally smooth gristle on the joint surface. This layer is called the _____

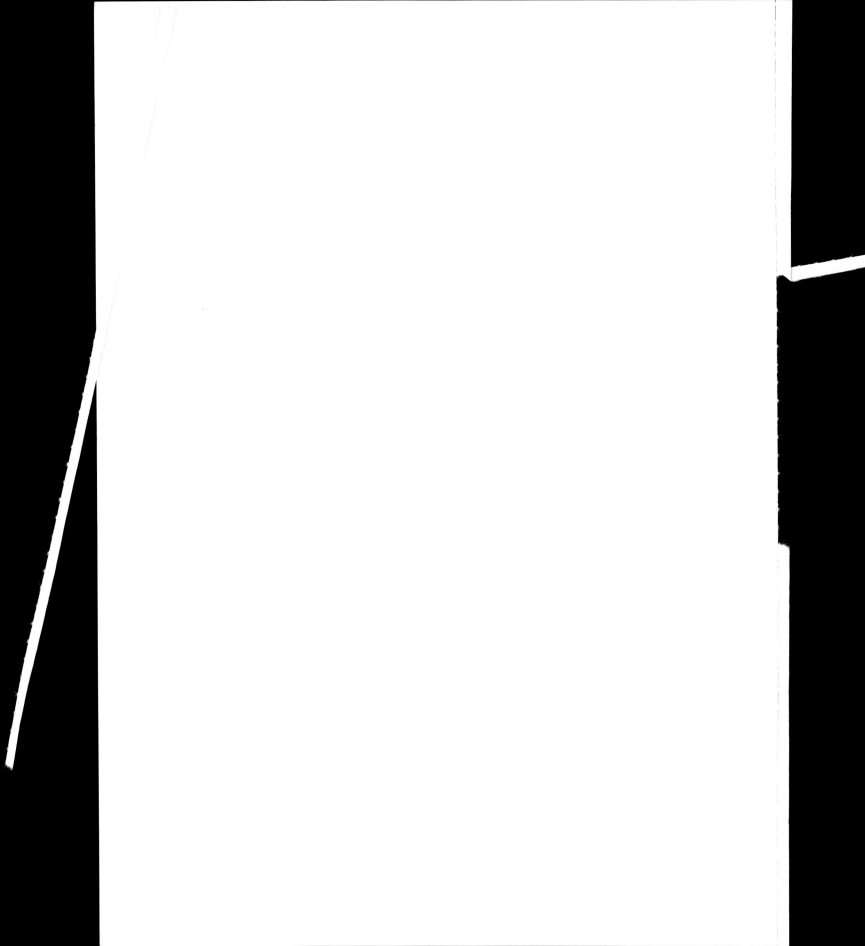

7

The Muscular System

I. Overview

The muscular system is composed of some 600 individual muscles, each of which is a distinct organ. Muscles usually work in groups to execute a body movement. The muscle that produces a given movement is called the **prime mover**; the muscle that produces the opposite action is the **antagonist**. There are three basic types of muscle tissue: **skeletal**, **smooth**, and **cardiac**. The focus of this chapter is skeletal muscle, which is attached to bones. Skeletal muscle is also called voluntary muscle, because normally it is under the conscious control of the will.

Skeletal muscles are activated by electrical impulses from the nervous system. A nerve fiber makes contact with a muscle cell at the **neuromuscular junction**. From this point, the impulse spreads along the muscle cell membrane, producing an electrical change called the **action potential**. As a result of this electrical change in the cells, the muscle can contract (shorten) to produce movement.

Muscle contraction occurs by the sliding together of protein filaments called **actin** and **myosin** within the cell. These filaments make contact only in the presence of calcium, which is released from the endoplasmic reticulum of the muscle cell when the action potential spreads along the cell membrane. **ATP** is the direct source of energy for the contraction. In order to manufacture ATP the cell must have adequate supplies of glucose and oxygen delivered by the blood. A reserve supply of glucose is stored in muscle cells in the form of a compound called **glycogen**, and additional oxygen is stored by a pigment in the cells called **myoglobin**.

When muscles do not receive enough oxygen, as during strenuous activity, they can produce a small amount of ATP and continue to function for a short period. As a result, however, the cells produce lactic acid which eventually causes muscle fatigue. The individual must then rest and continue to breathe in oxygen, which is used to convert the lactic acid into other substances. The amount of oxygen needed for this purpose is referred to as the **oxygen debt**.

Muscles act with the bones of the skeleton as **lever systems**, in which the joint is the pivot point or **fulcrum**. Exercise and proper body mechanics help in maintaining muscle health and effectiveness. Continued activity delays the undesirable effects of aging.

II. Topics for Review
A. General characteristics of skeletal muscles
B. The mechanism of muscle contraction
C. Muscle attachments
D. Muscle movement
E. Muscles of the head and the neck
F. Muscles of the upper extremities
G. Muscles of the trunk
H. Muscles of the lower extremities
I. Muscle metabolism during exercise
J. Body mechanics

III. Matching Exercises
Matching only within each group, print the answers in the spaces provided.

Group A
action potential contractility excitability
neuromuscular junction isotonic tonus
isometric

1. The capacity of a muscle to respond to a stimulus is known as _____

2. The point where a motor nerve fiber contacts a muscle cell is called the _____

3. Following stimulation of a muscle cell, the electrical change transmitted along the cell membrane is called a(n) _____

4. The capacity of a muscle fiber to undergo shortening is called _____

5. The normal partially contracted state of muscles is called _____

6. Muscle contractions in which the tone remains constant while the muscle shortens are _____

7. Those contractions in which there is a great increase in muscle tension without change in muscle length are called _____

Group B
glycogen actin lactic acid
calcium myoglobin oxygen
ATP

1. The substance that accumulates in muscles working without enough oxygen is _____

2. The ion that must be released into the muscle cell before contraction is _____

3. The immediate source of energy for muscle contraction is a substance called _____

4. The compound that stores glucose in muscle cells is _____

5. A protein filament needed to produce contraction in muscle cells is _____

6. During vigorous exercise muscles build up a need for _____

7. The compound that stores oxygen in muscle cells is _____

Group C

vasodilation   myosin   prime mover
antagonist   origin   insertion

1. The muscle that produces a given movement is the _____

2. The end of a muscle that puts a body part into action is the _____

3. A need for oxygen in muscle tissue produces a change in the blood vessels that brings more blood to the tissues. This change is called _____

4. The end of a muscle attached to a more fixed part of the body is the _____

5. A protein needed for contraction in muscle cells is _____

6. The muscle that must relax during a given movement is the _____

Group D

biceps brachii   pectoralis major   deltoid
latissimus dorsi   triceps brachii   trapezius
sternocleidomastoids   axilla

1. Working together, the two muscles on either side of the neck flex the head on the chest. They are the _____

2. Movement of the shoulder is a function of the _____

3. A powerful extensor of the arm (at the shoulder) used in swimming is the _____

4. The muscle capping the shoulder and upper arm, often used as an injection site, is the _____

5. A muscle on the front of the arm acts as a flexor of the elbow and a supinator of the hand. It is the _____

6. The large muscle on the back of the arm extends the elbow, as when delivering a blow (in boxing). Since it has three origins, it is called the _____

7. The large muscle of the upper chest flexes the arm across the body; it is called the _____

8. The pectoralis major and the latissimus dorsi both form part of the walls of the _____

Group E

gastrocnemius	levator ani	epimysium
aponeurosis	tendons	diaphragm
torticollis		

1. Muscles may be attached to bone by cordlike structures called _____

2. Some muscles are attached to bone by a sheet of _____

3. The connective sheath enclosing an entire muscle is the _____

4. The chief muscle of respiration is the _____

5. The chief muscle of the calf of the leg is the _____

6. The muscle of the pelvic floor that aids in defecation is the _____

7. Injury or spasm of a sternomastoid muscle may cause a condition called _____

Group F

sacrospinalis	intercostals	gluteus maximus
buccinator	sartorius	quadriceps femoris
iliopsoas		

1. Located between the ribs are muscles that aid in respiration, called the _____

2. The muscle that forms the fleshy part of the cheek is the _____

3. The longest muscle of the spine is the _____

4. Much of the fleshy part of the buttock is formed by the _____

5. The powerful flexor of the thigh is the _____

6. The muscle that extends the knee, as in kicking a ball, is the _____

7. The thin muscle that travels down and across the medial surface of the thigh is the _____

IV. Labeling

For each labeled muscle in the following drawings print the name on the appropriate numbered lines.

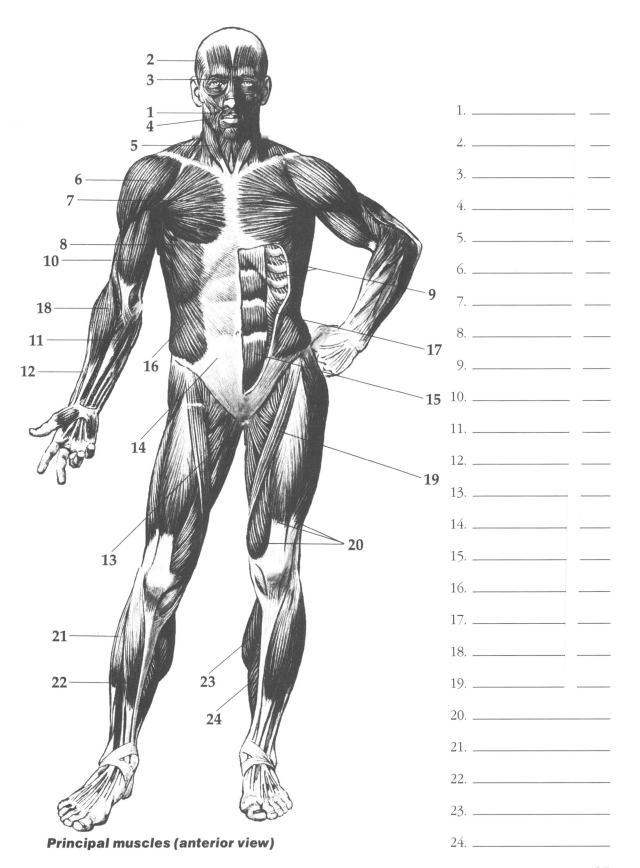

Principal muscles (anterior view)

1. _____ _____
2. _____ _____
3. _____ _____
4. _____ _____
5. _____ _____
6. _____ _____
7. _____ _____
8. _____ _____
9. _____ _____
10. _____ _____
11. _____ _____
12. _____ _____
13. _____ _____
14. _____ _____
15. _____ _____
16. _____ _____
17. _____ _____
18. _____ _____
19. _____ _____
20. _____ _____
21. _____ _____
22. _____ _____
23. _____ _____
24. _____ _____

1. _____
2. _____
3. _____
4. _____
5. _____
6. _____
7. _____
8. _____
9. _____
10. _____
11. _____
12. _____
13. _____
14. _____
15. _____
16. _____
17. _____

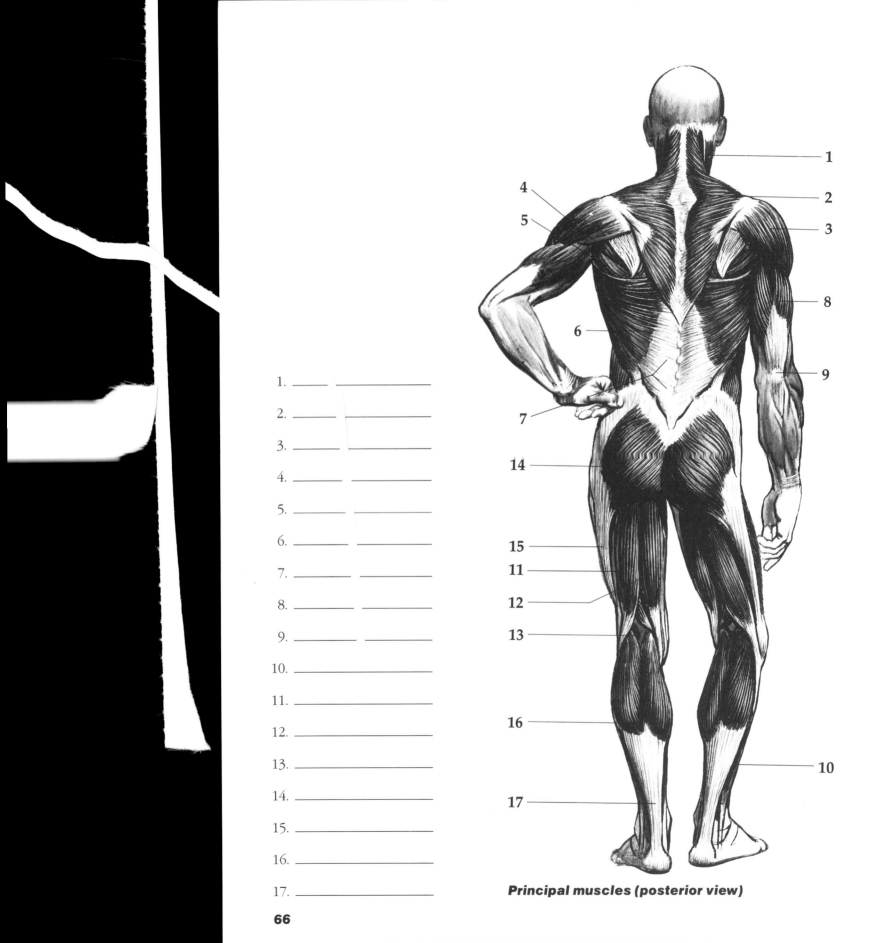

Principal muscles (posterior view)

66

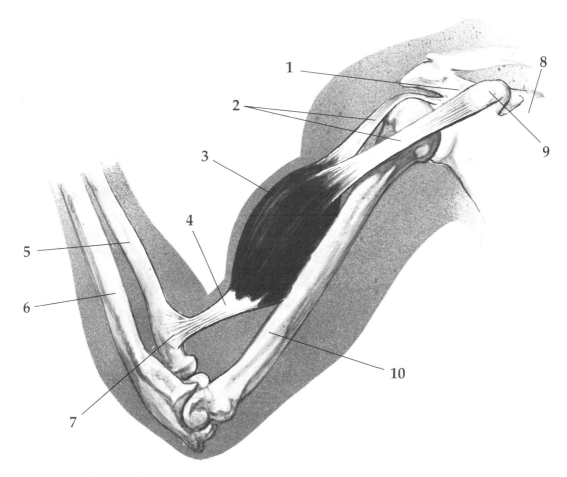

Tendons and muscles

1. _____
2. _____
3. _____
4. _____
5. _____

6. _____
7. _____
8. _____
9. _____
10. _____

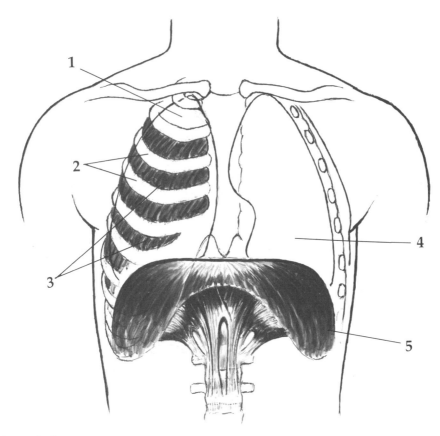

Location of diaphragm

1. _____

2. _____

3. _____

4. _____

5. _____

V. Completion Exercise

Print the word or phrase that completes each sentence.

1. Normally, muscles are in a partially contracted state, even though they are not in use at the time. This state of mild constant tension is called _____

2. A movement is initiated by a muscle or set of muscles called the _____

3. The end of a muscle that is attached to a part moved by that muscle is the _____

4. The muscle of the lips is the _____

5. Muscles functioning without enough oxygen fatigue as a result of the accumulation of _____

6. The muscle attachment that is usually relatively fixed is called its _____

7. The movement of a prime mover is opposed by a muscle or set of muscles called the _____

8. A group of muscles that covers the front and sides of the femur and extends the leg is the _____

9. There are four pairs of muscles for chewing. The muscle located at the angle of the jaw is called the _____

10. A superficial muscle of the neck and upper back acts on the shoulder. This muscle is the _____

11. The muscle on the front of the leg that raises the sole of the foot (dorsiflexion) is the _____

12. The largest forearm extensor is the _____

13. The muscular partition between the thoracic and abdominal cavities is the _____

14. The large fleshy muscle of the buttocks which extends the hip is the _____

15. The chief muscle of the calf is the toe dancer's muscle, named the _____

16. The band of connective tissue that attaches the gastrocnemius muscle to the heel is the _____

17. The muscle that turns the sole of the foot outward (eversion) is the _____

VI. Practical Applications

Study each discussion. Then print the appropriate word or phrase in the space provided.

Group A

Driver J and his three companions tried to race an oncoming train to an intersection. J misjudged the speed of the train, and the train crashed into the car. All four occupants of the car received multiple injuries.

1. Driver J was thrown against the steering wheel, which punctured his chest. This puncture involved the muscles between the ribs, called the _____

2. Mr. K, the occupant sitting next to the driver, suffered facial injuries, in which the muscle that encircles the eye was cut. This muscle is called the _____

3. Ms. L was thrown out of the car and received lacerations and fractures of the lower extremities, including the calf of the leg. The largest muscle of the leg is the _____

4. Mr. M received shoulder and upper back lacerations. They involved the muscle that covers the shoulder and abducts the arm, the _____

Group B

In the physical therapy department several patients were receiving physical therapy for muscle injuries.

1. Mrs. K had suffered a stroke that involved the left lower extremity. One of the large muscles used in standing forms most of the buttock, and is named the _____

2. Mr. P had suffered a fracture of the humerus and was receiving treatment for the damage to the large extensor of the elbow, located on the dorsal part of the arm. This muscle is the _____

3. Ms. L had been in a cast for a number of weeks, so she was receiving exercises for the strengthening of many body muscles, including the large muscle that originates from the middle and lower back and inserts on the arm bone (humerus). This strong swimming muscle is the _____

4. Ms. R, age 76, came in for exercises to strengthen some of her extensor muscles to prevent the further development of curvature of the spine. The large extensor muscle of the back needed particular attention. This is the _____

8

The Nervous System

I. Overview

The nervous system is the body's coordinating system, receiving, sorting out, and responding to both internal and external stimuli. The mechanism by which these activities occur is the **nerve impulse**, an electrical current which spreads along the membrane of the nerve cell or **neuron**. Each neuron is composed of a cell body and nerve fibers which extend from the cell body. **Dendrites** are fibers which carry impulses toward the cell body, and **axons** are fibers which carry impulses away from the cell body. Some axons are covered with a sheath of fatty material called **myelin** which insulates the fiber and speeds conduction along the fiber. Nerve cells make contact at junctions called **synapses**; the nerve impulse travels across the synapse by means of chemicals referred to as **neurotransmitters**. A neuron may be classified as either a sensory (afferent) type, which carries impulses toward the central nervous system, or a motor (efferent) type, which carries impulses away from the central nervous system. There are also connecting neurons within the central nervous system.

The nervous system as a whole is divided into the **central nervous system**, made up of the brain and the spinal cord, and the **peripheral nervous system**, made up of the cranial and spinal nerves. The brain consists of the cerebral hemispheres, diencephalon, brain stem, and cerebellum, each with specific functions. The spinal cord carries impulses to and from the brain. It is also a center for simple reflex activities in which responses are coordinated within the cord without traveling to the brain.

The brain and spinal cord are covered with three layers of fibrous membranes, the **meninges**. Also protecting these central structures is the **cerebrospinal fluid** produced in the ventricles of the brain.

Peripheral nerves, including the 12 pairs of cranial nerves and the 31 pairs of spinal nerves, connect all parts of the body with the central nervous system. Certain nerves within this system are grouped together as the **autonomic nervous system** which controls activities that go on more or less automatically. This system regulates the actions of glands, smooth muscle, and the heart. The autonomic nervous system has two divisions, the **sympathetic** and **parasympathetic** nervous systems, which have opposite effects on given organs.

II. Topics for Review

A. The function of nerve tissue
   1. The nerve cell and its fibers
   2. The nerve impulse
   3. Nerves
B. Divisions of the nervous system
   1. Central nervous system
   2. Peripheral nervous system
   3. Autonomic nervous system
C. Central nervous system
   1. Brain
      a. Cerebral hemispheres
      b. Diencephalon
      c. Brain stem
      d. Cerebellum
   2. Spinal cord
      a. Structure
      b. Function
   3. Coverings of the brain and spinal cord
   4. Cerebrospinal fluid
D. Peripheral nervous system
   1. Cranial nerves
   2. Spinal nerves
E. Autonomic nervous system
   1. Structure
   2. Functions

III. Matching Exercises

Matching only within each group, print the answers in the spaces provided

Group A

brain and spinal cord	brain stem	nerve
autonomic nervous system	peripheral nervous system	coordination
stimuli	cerebral hemispheres	

1. The main function of the nervous system is _____

2. The internal and external changes that affect the nervous system are called _____

3. For study purposes, the entire nervous system has been divided into two large systems. One of these, the central nervous system, is composed of the _____

4. The cranial and spinal nerves constitute the _____

5. The cerebrum is the largest part of the brain. It is divided into right and left parts called the _____

6. The midbrain, pons, and medulla oblongata form the _____

7. The peripheral nerves that regulate activities going on more or less automatically are grouped together as the _____

8. Impulses are conducted from one place to another by a bundle of nerve fibers, the _____

Group B

neuron	nerve impulse	mixed
reflex	nerve fibers	autonomic nervous system
neurilemma	dendrite	synapse

1. Most nerves contain both afferent and efferent fibers and are thus described as _____

2. The nerve cell, including the cell body and its projections, is a(n) _____

3. The sympathetic and parasympathetic nervous systems are the two functionally opposing parts of the _____

4. The threadlike cytoplasmic projections of the nerve cell are the _____

5. An electrical charge which spreads along the membrane of a nerve cell is the _____

6. A nerve fiber that carries impulses toward the cell body is a(n) _____

7. A nervous response that is coordinated within the spinal cord is termed a(n) _____

8. The point of junction between two neurons is a(n) _____

9. A sheath around some axons that aids in regeneration is the _____

Group C

fissure	lobes	afferent
receptor	gyri	motor
cerebral cortex	ventricles	neurotransmitter

1. The point at which a stimulus is received is called the _____

2. A chemical that carries the nerve impulse across a synapse is a(n) _____

3. Impulses must be carried to and away from the brain and spinal cord. Neurons that conduct impulses to the brain and cord are described as _____

4. Nerves that carry impulses away from the brain and cord to muscles and glands are classified as _____

5. All thought, association, and judgment take place in the _____

6. The deep groove that divides the main part of the cerebrum into two hemispheres is the longitudinal _____

7. The sulci serve to separate the gray matter into elevated portions known as _____

8. Cerebrospinal fluid is produced in spaces within the brain called _____

9. Each hemisphere of the brain is divided into regions, each of which regulates certain types of functions. These areas are called _____

Group D

corpus callosum   motor cortex   parietal lobe
occipital lobe   meninges   myelin
temporal lobe   thalamus

1. In the disorder known as multiple sclerosis there is degeneration of the fatlike substance that covers many nerve fibers. This sheath is composed of _____

2. The three brain coverings are collectively known as the _____

3. In each frontal lobe is an area that controls voluntary muscles. This is the _____

4. Pain, touch, and temperature are interpreted in the sensory area which is contained in the _____

5. Impulses received by the ear are interpreted in the auditory center, which is located in the _____

6. Messages from the retina are interpreted in the visual area of the _____

7. A band of white matter which acts as a bridge between the cerebral hemispheres is the _____

8. Two masses of gray matter which are located in the diencephalon act as relay centers monitoring sensory stimuli. These two masses constitute the _____

Group E

cerebrum   cerebellum   medulla oblongata
blood pressure   diencephalon   hypothalamus
corpora quadrigemina

1. The part of the brain that contains the thalamus and the hypothalamus is the _____

2. The portion of the brain that coordinates voluntary muscles and helps to maintain balance is the _____

3. The two cerebral hemispheres form much of the largest part of the brain, the _____

4. The respiratory, cardiac, and vasomotor centers are found in the _____

5. The vasomotor center affects muscles in the blood vessel walls and thus influences _____

6. Body temperature, sleep, the heartbeat, and water balance are among the vital body functions regulated by the _____

7. The relay centers for eye and ear reflexes are located in the midbrain. They are the four _____

Group F

afferent nerves   white matter   efferent nerves
receptor   effector   dorsal root

1. Myelinated fibers make up the regions of the central nervous system described as _____

2. The cell bodies of sensory neurons are located in a ganglion on the _____

3. The spinal cord has several essential functions. One of these is to conduct sensory impulses upward to the brain in tracts within the cord. These impulses are brought to the cord by _____

4. The spinal cord also functions as a pathway for conducting motor impulses from the brain downward in descending tracts. These motor impulses leave the cord by way of _____

5. The reflex pathway begins with the part of a sensory neuron called a(n) _____

6. The sensory neuron conducts an impulse to a central neuron which then transfers it to a motor neuron. This typical reflex pathway terminates in a gland or a muscle termed a(n) _____

Group G

dura mater   arachnoid membrane   pia mater
arachnoid villi   choroid plexuses   subarachnoid space

1. The innermost layer of the meninges, the delicate connective tissue membrane in which there are many blood vessels, is the _____

2. The weblike middle meningeal layer is the _____

3. The outermost meningeal layer, which is the thickest and toughest, is also made of connective tissue. It is the _____

4. Normally, the cerebrospinal fluid helps protect the brain and spinal cord against shock. This fluid is formed inside the brain ventricles by the _____

5. Normally, the fluid flows freely from ventricle to ventricle and finally out into the _____

6. The fluid is returned to the blood in the venous sinuses through the projections called _____

Group H

visual area   auditory speech center   visual speech center
written speech center   sensory area   left cerebrum

1. Pain, touch, temperature, size, and shape are interpreted in the parietal lobe, in a section called the _____

2. The understanding of words takes place with the development of a temporal lobe area known as the _____

3. The muscles in the right side of the body are controlled by the _____

4. Messages from the retina are interpreted in the region of the occipital lobe known as the _____

5. The ability to read with understanding comes with the development of the _____

6. The ability to write words, which usually is a late phase in a person's total language comprehension, is a function of the _____

Group I

ganglion   plexuses   brachial plexus
cervical plexus   roots   visceral nervous system
somatic nervous system

1. Each spinal nerve is attached to the spinal cord by branches called _____

2. Involuntary control over smooth muscles, glands, and the heart is brought about by the _____

3. A collection of nerve cell bodies usually found outside the central nervous system is a(n) _____

4. A short distance away from the spinal cord, each spinal nerve branches into two divisions; the branches of the larger division interlace to form _____

5. The shoulder, the arm, the wrist, and the hand are supplied by branches from the _____

6. Motor impulses to the neck muscles are supplied by the _____

7. Skeletal muscles are controlled by the _____

Group J

parasympathetic nervous system
sympathetic nervous system
hypoglossal nerve
olfactory nerve
vagus nerve
vestibulocochlear nerve
optic nerve
facial nerve
oculomotor nerve
trigeminal nerve

1. Recall the functions of the autonomic nervous system. The part that acts to prepare the body for emergency situations is the _____

2. The part of the autonomic nervous system that aids the digestive process is the _____

3. Impulses controlling tongue muscles are carried by the _____

4. General sense impulses from the face and head are carried through the three branches of the _____

5. Sense fibers for hearing are contained within the _____

6. The muscles of facial expression are supplied by branches of the _____

7. The nerve that carries smell impulses to the brain is the _____

8. The contraction of most eye muscles is controlled by the _____

9. The sensory nerve that carries visual impulses is the _____

10. Most of the organs in the thoracic and abdominal cavities are supplied by the _____

IV. Labeling

For each of the following illustrations, print the name or names of each labeled part on the numbered lines.

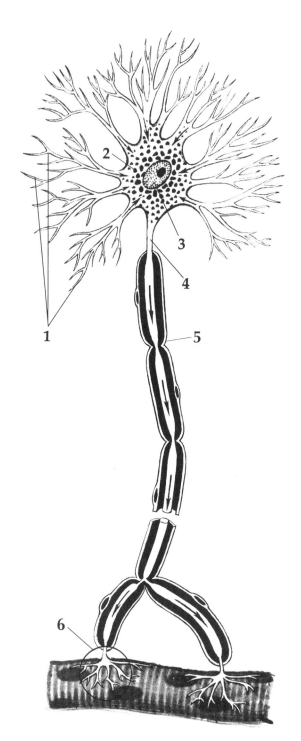

Diagram of a motor neuron

1. _____   4. _____

2. _____   5. _____

3. _____   6. _____

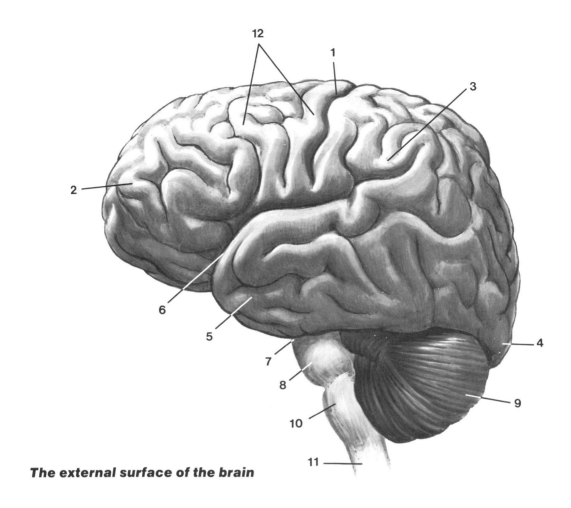

The external surface of the brain

1. _____
2. _____
3. _____
4. _____
5. _____
6. _____
7. _____
8. _____
9. _____
10. _____
11. _____
12. _____

79

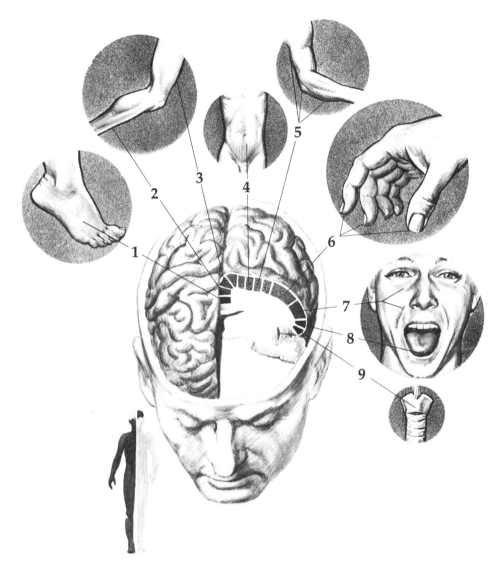

The motor area of the left cerebral hemisphere

1. _____
2. _____
3. _____
4. _____
5. _____
6. _____
7. _____
8. _____
9. _____

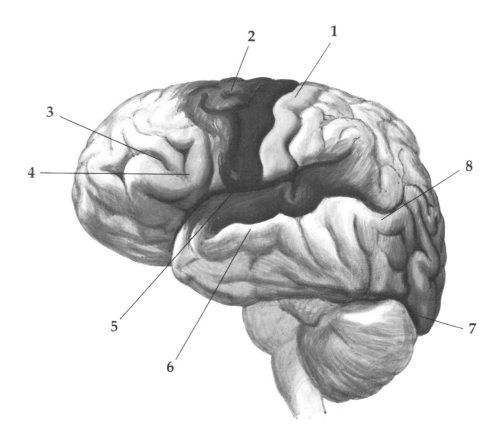

The functional areas of the cerebrum

1. _____
2. _____
3. _____
4. _____
5. _____
6. _____
7. _____
8. _____

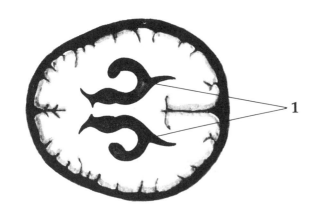

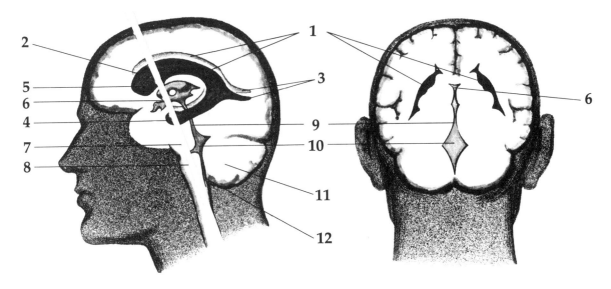

Brain ventricles

1. _____   7. _____

2. _____   8. _____

3. _____   9. _____

4. _____   10. _____

5. _____   11. _____

6. _____   12. _____

1. _____
2. _____
3. _____
4. _____
5. _____
6. _____
7. _____
8. _____
9. _____
10. _____
11. _____
12. _____
13. _____
14. _____
15. _____

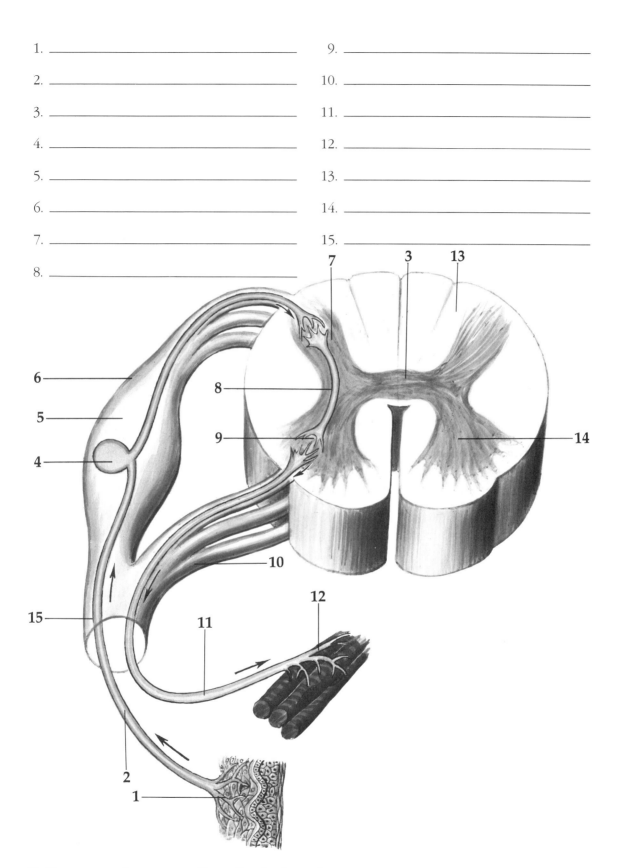

Reflex arc and cross section of spinal cord

83

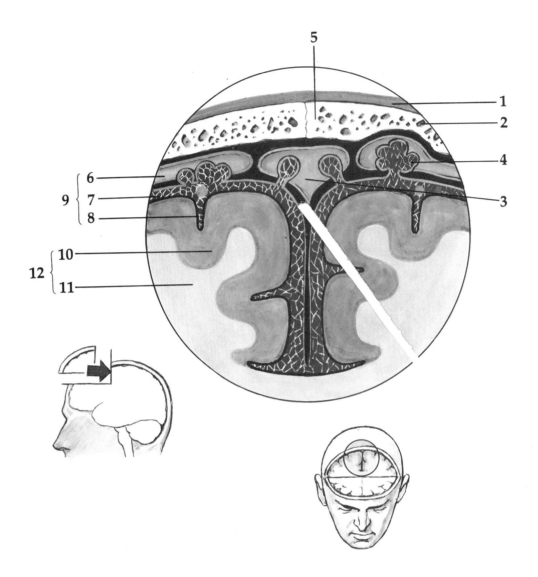

Frontal (coronal) section of top of head to show meninges and related parts

1. _____
2. _____
3. _____
4. _____
5. _____
6. _____
7. _____
8. _____
9. _____
10. _____
11. _____
12. _____

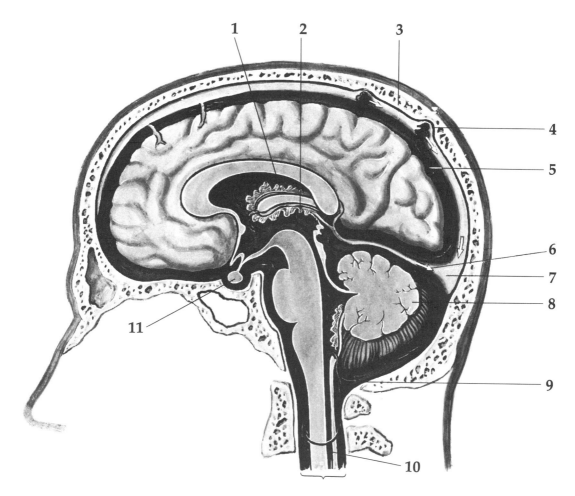

Flow of cerebrospinal fluid

1. _____
2. _____
3. _____
4. _____
5. _____
6. _____
7. _____
8. _____
9. _____
10. _____
11. _____

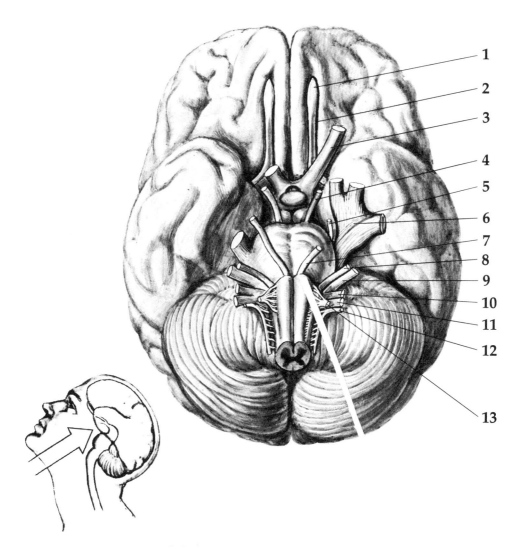

Base of brain, showing cranial nerves

1. _____
2. _____
3. _____
4. _____
5. _____
6. _____
7. _____
8. _____
9. _____
10. _____
11. _____
12. _____
13. _____

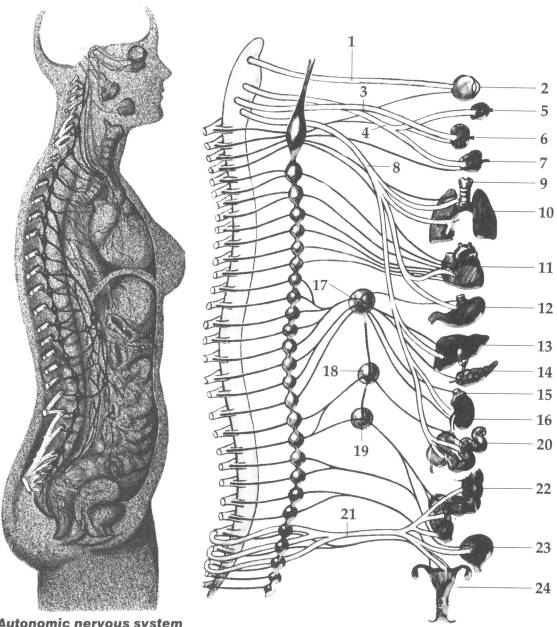

Autonomic nervous system

1. _____
2. _____
3. _____
4. _____
5. _____
6. _____
7. _____
8. _____
9. _____
10. _____
11. _____
12. _____
13. _____
14. _____
15. _____
16. _____
17. _____
18. _____
19. _____
20. _____
21. _____
22. _____
23. _____
24. _____

87

V. Completion Exercise

Print the word or phrase that correctly completes each sentence.

1. The brain and spinal cord together are usually referred to as the _____

2. The cranial and spinal nerves together form the part of the nervous system described as the _____

3. Activities of the body that go on automatically are under control of the _____

4. A nerve cell is also called a(n) _____

5. A nerve fiber that conducts impulses away from the cell body is a(n) _____

6. A specialized nerve ending that can detect a stimulus is a(n) _____

7. The largest branch of the lumbrosacral plexus is the _____

8. The slightly curved groove or depression along the side of the brain which separates the temporal lobe from the rest of the cerebral hemisphere is the _____

9. The fluid-filled spaces within the cerebral hemispheres are the _____

10. Through a large opening in the base of the skull, the spinal cord connects with a part of the brain called the _____

11. Collections of cell bodies within the central nervous system are called _____

12. Unmyelinated fibers and cell bodies make up the portions of the central nervous system described as the _____

VI. Practical Applications

Study each discussion. Then print the appropriate word or phrase in the space provided.

1. Mr. H, age 42, had been suffering for several weeks from persistent headaches. As part of the diagnostic study, some of the fluid was removed from the ventricles in Mr. H's brain and replaced with air. This fluid is the _____

2. As a result of the various studies done in Mr. H's case, it was determined that a tumor was present in the left lateral ventricle. Surrounding the left ventricle is the _____

3. Ms. S's symptoms included paralysis and various motor disturbances. The diagnosis of myelitis, or inflammation of the spinal cord, was made. This nerve cord is located in a space called the _____

4. Young A, age 10, was brought to the emergency clinic following a bicycle accident in which he received trauma to his head. Diagnostic studies included a CT (computed tomography) scan of the head. The purpose of this study was to detect any tears in the brain coverings such that blood could collect in the space between the brain and the skull. The tough outer covering of the brain and cord is called the _____

9

The Sensory System

I. Overview

Through the functioning of the **sensory receptors**, we are made aware of changes taking place both internally (within the body) and externally (outside the body). Any change that produces a response in the nervous system is termed a **stimulus**.

The **special senses**, so-called because the receptors are limited to a relatively small area of the body, include the vision sense, the hearing sense, and the senses of taste and smell. The receptors of the eye are the **rods and cones** located in the retina. The hearing receptors are found in a portion of the inner ear called the **cochlea**. Receptors for the chemical senses of taste and smell are located on the tongue and in the upper part of the nose respectively.

The **general senses** are scattered throughout the body; they have to do with pressure, temperature, pain, touch, and position. Receptors for the sense of position, known as **proprioceptors**, are found in muscles, tendons, and joints.

The nerve impulses generated in a receptor cell by a stimulus must be carried to the central nervous system by way of a sensory (afferent) neuron. Here the information is processed and a suitable response is made.

II. Topics for Review

A. The eye
 1. Protective structures of eyeball
 2. Coats of eyeball
 3. Pathway of light rays
 4. Sensory receptors
 5. Extrinsic eyeball muscles
 6. Intrinsic eyeball muscles
 7. Nerve supply to the eye
 8. Lacrimal apparatus

B. The ear
   1. External ear; pinna and auditory canal
   2. Middle ear
   3. Internal ear
C. Other organs of special sense
   1. Taste receptors
   2. Smell receptors
D. General senses
   1. Pressure
   2. Temperature
   3. Touch
   4. Pain
   5. Position

III. Matching Exercises

Matching only within each group, print the answers in the spaces provided. The same answer may be used more than once.

Group A

cornea   aqueous humor   choroid coat
accommodation   rods and cones   color
retina   vitreous body

1. The innermost coat of the eyeball, the nerve tissue layer, includes the receptors for the sense of vision. This structure is the _____

2. The pigmented middle tunic of the eyeball is the vascular _____

3. Light rays pass through a series of transparent eye parts. The outermost of these is the _____

4. The watery fluid that fills much of the eyeball in front of the crystalline lens and also helps to maintain the slight curve in the cornea is the _____

5. The spherical shape of the eyeball is maintained by a jellylike material located behind the crystalline lens. This is the _____

6. The receptors for the sense of vision are called the _____

7. There are three types of cones, each of which is sensitive to a different _____

8. The elacticity of the lens enables it to become thicker and bend the light rays as necessary for near vision. This process is _____

Group B

iris   pupil   ciliary body
media   sclera   receptors
optic disk   conjunctiva

1. The opaque outermost layer of the eyeball is made of firm, tough connective tissue. This coat is the _____

2. The central opening in the iris contracts or dilates according to need. This opening is the _____

3. The crystalline lens is one of the transparent refracting parts of the eye. Collectively they are called _____

4. The rods and cones of the retina are the visual _____

5. The membrane that lines the eyelids is the _____

6. The region of connection between the optic nerve and the eyeball is lacking in rods and cones and is commonly called the blind spot. Another term for this is _____

7. The shape of the lens is altered by the muscle of the _____

8. The pupil is the central opening in the colored part of the eye, the _____

Group C

fovea centralis   iris   refraction
sphincter   lacrimal gland   intrinsic
extrinsic

1. The muscles that are attached to bones of the orbit and to the sclera are located outside the eyeball and are described as _____

2. When a light is flashed in the eye the pupil is reduced in size owing to the contraction of a circular iris muscle which forms a(n) _____

3. The amount of light entering the eye is controlled by the _____

4. The process of bending which makes it possible for light from a large area to be focused on a small surface is known as _____

5. Tears serve an important protective function for the eye. They are produced by the _____

6. The clearest point of vision is a depressed area in the retina, the _____

7. The muscles of the iris and ciliary body are located entirely within the eyeball and so are described as _____

Group D

oval window   external auditory canal   eustachian tube
ossicles   endolymph   tympanic membrane
perilymph   pinna   mastoid air cells

1. Located at the end of the auditory canal is the eardrum, or _____

2. The three small bones within the middle ear cavity are the _____

3. The spaces within the temporal bone which connect with the middle ear cavity through an opening are called the _____

4. Sound waves are conducted to the fluid of the internal ear by vibrations of the membrane that covers the _____

5. Air is brought to the middle ear cavity by means of the auditory tube which is also called the _____

6. The fluid of the inner ear contained within the bony labyrinth and surrounding the membranous labyrinth is called _____

7. The fluid contained within the membranous labyrinth is called _____

8. Sound waves enter the _____

9. Another name for the projecting part, or auricle, of the ear is the _____

Group E

optic nerve   vestibule   ophthalmic nerve
oculomotor nerve   cochlear duct   cochlear nerve
rods   equilibrium

1. The organ of hearing is made up of receptors located in the _____

2. The branch of the vestibulocochlear nerve that carries hearing impulses is the _____

3. The entrance area that communicates with the cochlea and that is next to the oval window is the _____

4. Visual impulses received by the rods and cones of the retina are carried to the brain by the _____

5. Impulses of pain, touch, and temperature are carried to the brain by a branch of the fifth cranial nerve, the _____

6. The largest cranial nerve carrying motor fibers to the eyeball muscles is the _____

7. The retinal receptors that function in dim light are the _____

8. The semicircular canals and the vestibule contain receptors for the sense of _____

Group F

taste buds
adaptation
proprioceptors

ceruminous
vitamin A

olfactory
pressure

1. The sense of taste involves two cranial nerves as well as receptors known as _____

2. Night blindness may result from a deficiency of _____

3. Among the general senses is that concerned with deep sensibility, commonly called the sense of _____

4. In the case of many sensory receptors, including those for temperature, the receptors adjust themselves so that one does not feel the sensation so acutely if the original stimulus is continued. Such an adjustment to the environment is called _____

5. The wax glands located in the external auditory canal are described as _____

6. The pathway for impulses from smell receptors is the first cranial nerve, the _____

7. Receptors that transmit information on the position of body parts are called _____

IV. Labeling

For each of the following illustrations print the name or names of each labeled part on the numbered lines.

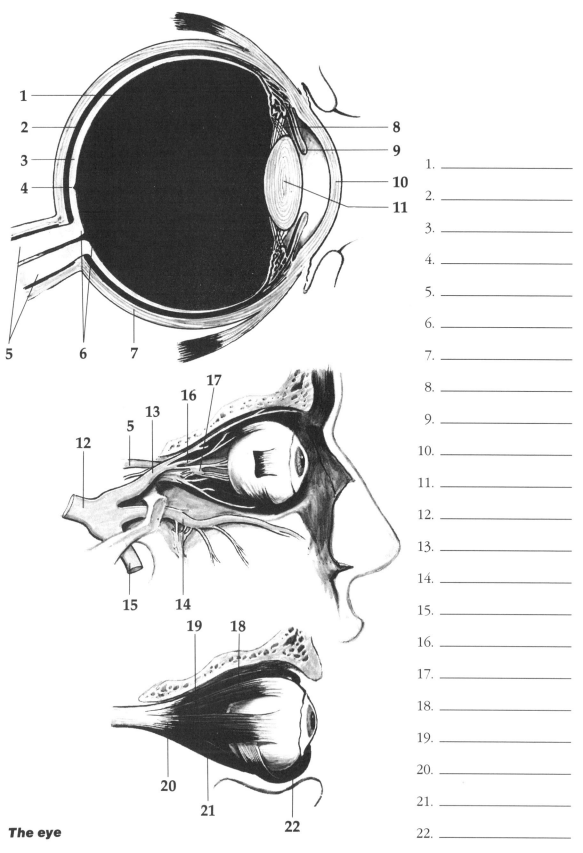

The eye

1. _____
2. _____
3. _____
4. _____
5. _____
6. _____
7. _____
8. _____
9. _____
10. _____
11. _____
12. _____
13. _____
14. _____
15. _____
16. _____
17. _____
18. _____
19. _____
20. _____
21. _____
22. _____

1. _____
2. _____
3. _____
4. _____
5. _____
6. _____
7. _____
8. _____
9. _____
10. _____
11. _____
12. _____
13. _____
14. _____
15. _____
16. _____
17. _____
18. _____
19. _____
20. _____

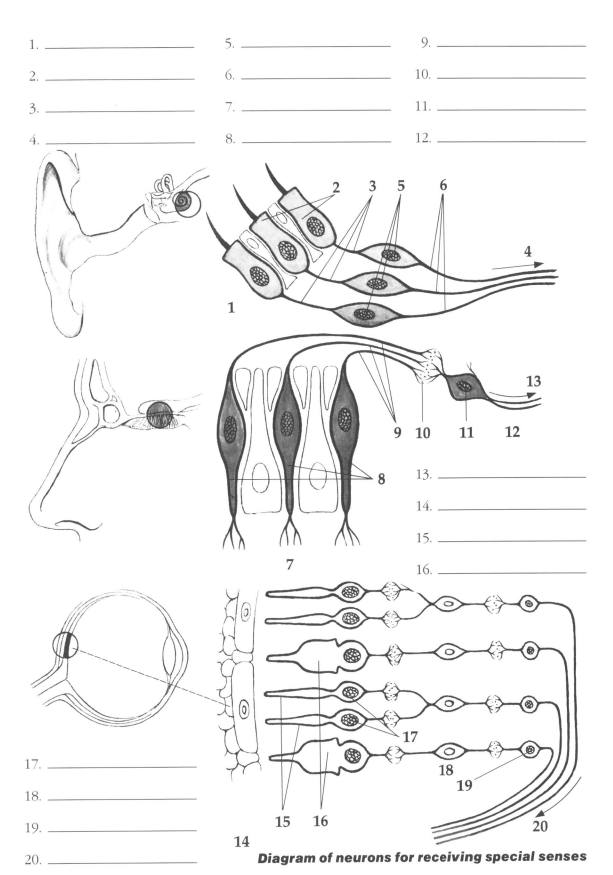

Diagram of neurons for receiving special senses

97

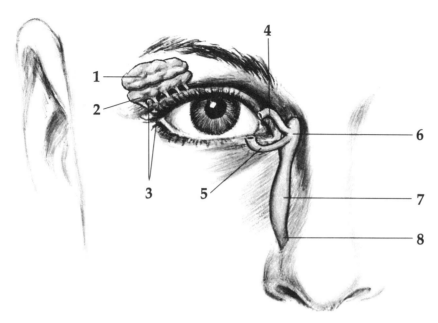

Lacrimal apparatus

1. _____
2. _____
3. _____
4. _____
5. _____
6. _____
7. _____
8. _____

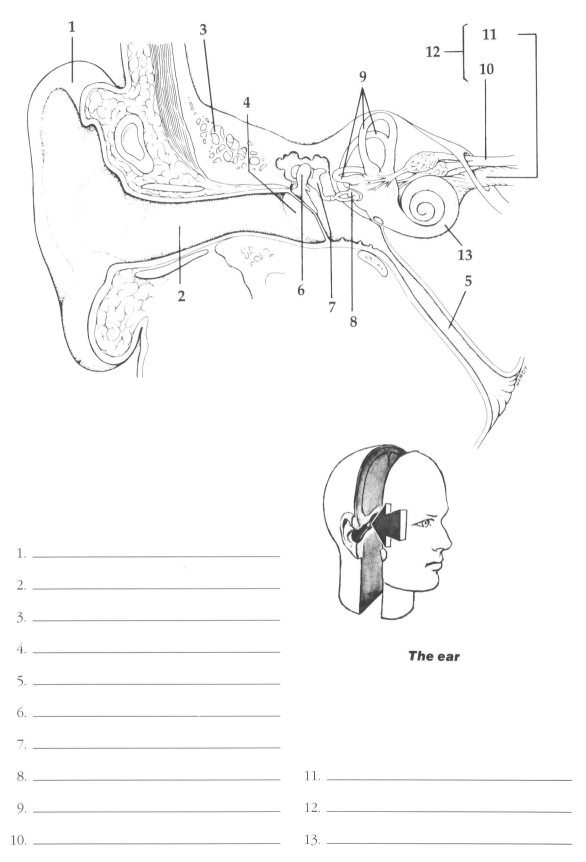

The ear

1. _____
2. _____
3. _____
4. _____
5. _____
6. _____
7. _____
8. _____
9. _____
10. _____
11. _____
12. _____
13. _____

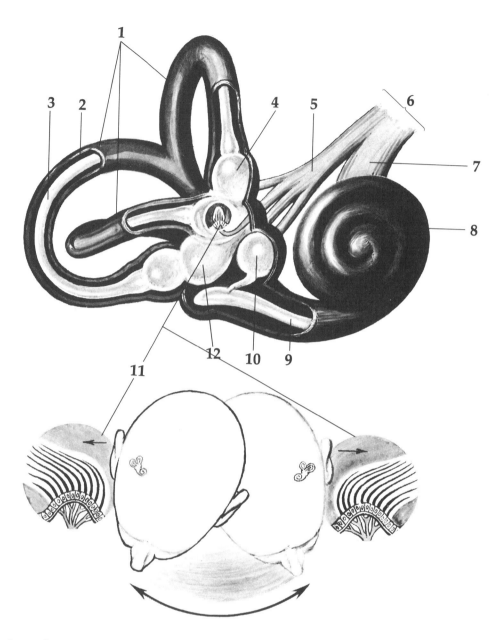

The internal ear

1. _____
2. _____
3. _____
4. _____
5. _____
6. _____
7. _____
8. _____
9. _____
10. _____
11. _____
12. _____

V. Completion Exercise

Print the word or phrase that correctly completes each sentence.

1. The nerve fibers of the vestibular and cochlear nerves join to form the nerve called the _____

2. The inner ear spaces contain fluids involved in the transmission of sound waves. The one that is inside the membranous cochlea and that stimulates the receptors is the _____

3. The taste receptors of the tongue are located along the edges of small depressed areas, or _____

4. The nerves involved in the sense of taste are the facial and the _____

5. Pain that is felt in an outer part of the body such as the skin, yet that originates internally near the area where it is felt, is called _____

6. The very widely distributed free nerve endings are the receptors for the most important protective sense, namely that for _____

7. The tactile corpuscles are the receptors for the sense of _____

8. The nerve endings that relay impulses which aid in judging position and changes in location of parts with respect to each other are the _____

9. The sense of position is partially governed by several structures in the internal ear, including two small sacs and the three membranous _____

10. When you enter a darkened room, it takes a while for the rods to begin to function. This interval is known as the period of _____

VI. Practical Applications

Study each discussion. Then print the appropriate word or phrase in the space provided.

Group A

While observing in the emergency ward the student nurse noted the following cases.

1. Ten-year-old K had been riding his bicycle while he threw glass bottles to the sidewalk. A fragment of glass flew into one eye. Examination at the hospital showed that there was a cut in the transparent window of the eye, the _____

2. On further examination of K, the colored part of the eye was seen to protrude from the wound. This part is the _____

3. K's treatment included antiseptics, anesthetics, and suturing of the wound. Medication was instilled in the saclike structure at the front of the eyeball. This sac is lined with a thin epithelial membrane, the _____

4. A construction worker, Mr. J, was admitted because of an accident in which a piece of steel penetrated his eyeball and caused such an extensive wound that material from the inside of the eyeball oozed out. Mr. J tried to relieve the pain by forcing the jellylike material out through the wound at the front and side of his eyeball. This matter, which maintains the shape of the eyeball, is called the _____ __

Group B

An ear, nose, and throat specialist treated the following patients one morning.

1. Mrs. B complained of some deafness and a sense of fullness in her outer ear. Examination revealed that the wax in her ear canal had hardened and formed a plug of (*scientific name*) _____ __

2. Mr. J, age 72, complained of gradually increasing deafness although he had no symptoms of pain or other problems related to the ears. Examination revealed that his deafness was the type called nerve deafness. The cranial nerve that carries impulses related to hearing to the brain is called the _____ _

3. Baby L was brought in by his mother because he awakened crying, and was holding the right side of his head. He had been suffering from a cold but now he seemed to be in pain. Examination revealed a bulging red eardrum. The eardrum is also called the _____

4. Elderly Mr. N had a hearing loss due to destruction of the nerve endings located in the spiral-shaped part of the internal ear, the _____

10

The Endocrine System and Hormones

I. Overview

The endocrine glands are ductless glands that release their secretions directly into the bloodstream. These secretions, called **hormones**, are chemical messengers that regulate growth, metabolism, sexual development, and behavior. The endocrine system is composed of organs that have the secretion of hormones as a primary function. These include the following glands: pituitary (hypophysis), thyroid, parathyroids, adrenals, pancreas, gonads, thymus, and pineal.

The endocrine system and the nervous system are the main coordinating and controlling systems of the body. Both are activated, for example, in helping the body respond to stress. These two systems meet in the **hypothalamus**, a region of the diencephalon of the brain. The hypothalamus is directly above and connected to the **pituitary**. By means of nerve stimulation and hormones, the hypothalamus controls the two lobes of the pituitary. Hormones released from the pituitary, in turn, control the other endocrine glands. The other main mechanism for controlling hormone secretion is **negative feedback**, in which hormone levels, or substances released as a result of hormone action, serve to regulate the production of that hormone.

Chemically, hormones are either **proteins** or **steroids**. The cells on which hormones act make up the **target tissue**. These cells have **receptors** to which the hormone attaches. Either directly or by means of a second messenger, hormones affect the activity of the DNA within the cell and the manufacture of proteins. In this manner they regulate the activities of the cell.

In addition to the endocrine glands, some other structures, including the kidney, stomach, and small intestine, secrete hormones. **Prostaglandins** are hormone-like substances produced by cells throughout the body. They have a variety of effects and are currently under study.

II. Topics for Review
A. General characteristics of the endocrine system
B. Hormones
   1. Chemical makeup
   2. Method of action
   3. Regulation
C. The endocrine glands and their hormones
   1. Control of the pituitary by the hypothalamus
D. Other hormone-producing organs
E. Hormones and stress

III. Matching Exercises
Matching only within each group, print the answer in the space provided.

Group A

parathyroid glands   thyroid   islets of Langerhans
calcitonin   medulla   suprarenal glands
hormones

1. The substances produced by endocrine glands are known as   _____

2. The adrenal glands are also known as the   _____

3. The groups of hormone-secreting cells scattered throughout the pancreas are known as the   _____

4. The largest of the endocrine glands is located in the neck. It is the   _____

5. The inner part of the adrenal gland is called the   _____

6. Located behind the thyroid gland and embedded in its capsule are the four   _____

7. The hormone produced by the thyroid gland and active in calcium metabolism is   _____

Group B

DNA   proteins   hormones
thyroxine   negative feedback   target tissues
hypothalamus   second messenger

1. The body's "chemical messengers" are the   _____

2. The part of the brain that controls the pituitary gland is the   _____

3. All hormones except the sex hormones and the hormones of the adrenal cortex are classified chemically as   _____

4. Hormones work only on specific tissues known as the   _____

5. Production of heat and energy in the body tissues is regulated mainly by the hormone _____

6. Within the cell, hormones affect the working of the genetic material, the _____

7. If a hormone does not enter a cell it acts indirectly by means of a substance within the cell called a(n) _____

8. Most hormones are regulated by a self-controlling mechanism known as _____

Group C

iodine	insulin	pituitary
adrenal	parathyroid hormone	receptors

1. A target tissue responds to a given hormone because it has areas to which the hormone attaches. These areas are called _____

2. The endocrine gland that is divided into anterior and posterior lobes is the _____

3. The amount of calcium dissolved in the circulating blood is partly regulated by a secretion from small glands on the surface of the thyroid. This secretion is called _____

4. In order to provide for normal sugar utilization in the tissues, the islets of Langerhans must produce a hormone called _____

5. The endocrine gland composed of an external cortex and an internal medulla which act as separate glands with specific functions is the _____

6. In order that thyroxine may be manufactured, the blood must contain an adequate supply of _____

Group D

steroids	cortisol	calcium
insulin	placenta	anterior lobe
lymphocytes	oxytocin	glucagon

1. A portal system connects the hypothalamus to the region of the pituitary called the _____

2. In the disorder known as diabetes mellitus, sugar is not "burned" in the tissues to produce energy. This is due to a lack of the hormone _____

3. The normal development of the embryo is aided by hormones from the ovaries, pituitary, and an organ present only during pregnancy, namely the _____

4. The sex hormones and the hormones of the adrenal cortex are classified chemically as _____

5. Hydroxycholecalciferol, a hormone-like substance produced from vitamin D, regulates intestinal absorption of the mineral _____

6. During stressful situations, such as an injury or surgery, the body is protected somewhat by an adrenal hormone (a glucocorticoid) that acts to reduce inflammation. This hormone is usually called _____

7. The hormone produced by the islets of Langerhans that raises blood sugar levels is _____

8. The thymus produces hormones that stimulate the production of cells needed in the body's defenses against infection. These cells are the _____

9. The hormone from the posterior pituitary which causes uterine contraction is called _____

Group E

aldosterone   antidiuretic hormone   ACTH
cortisol   estrogen   progesterone
epinephrine   kidney

1. Blood pressure is raised and the rate of the heartbeat is increased by the chief hormone of the adrenal medulla, _____

2. Regulation of reabsorption of sodium and secretion of potassium in the kidney tubules is a function of the adrenal cortex hormone _____

3. Testosterone is produced by the male sex glands; the female sex glands produce a hormone that most nearly parallels testosterone in its action. This hormone is called _____

4. The hormone produced in the posterior lobe of the pituitary which regulates water reabsorption by the kidney is called _____

5. A hormone that is necessary for normal development of pregnancy is one produced by the female sex glands. It is called _____

6. When the needs of the body are such that amino acids must be changed to sugar instead of protein, the adrenal cortex produces large amounts of the hormone _____

7. The adrenal cortex is stimulated by the anterior pituitary hormone known as _____

8. The hormone erythropoietin, which stimulates production of red blood cells, is produced in the _____

IV. Labeling

Print the name or names of each labeled part on the numbered lines.

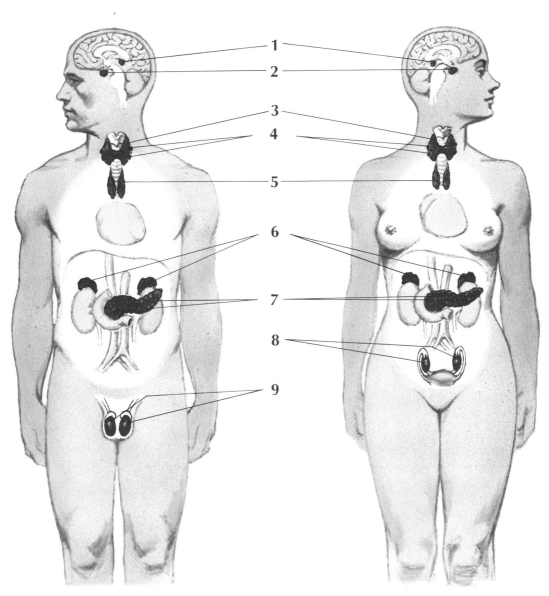

Glands of the endocrine system

1. _____

2. _____

3. _____

4. _____

5. _____

6. _____

7. _____

8. _____

9. _____

V. Completion Exercise

Print the word or phrase that correctly completes each sentence.

1. The region of the brain that controls the pituitary gland (hypophysis) is the _____

2. Growth hormone is produced by the region of the pituitary called the _____

3. The hypothalamus stimulates the anterior pituitary to produce ACTH, which in turn stimulates hormone production by the _____

4. Hormones secreted by the anterior lobe of the pituitary that control the activity of the sex glands or gonads are described as _____

5. Hormone-like substances that have a variety of effects, including the promotion of inflammation and the production of uterine contractions, are the _____

6. When the level of glucose in the blood decreases to less than average, the islet cells of the pancreas release less insulin. The result is an increase in blood glucose. This is an example of the mechanism called _____

VI. Practical Applications

Study each discussion. Then print the appropriate word or phrase in the space provided.

1. Mr. J, age 23, required evaluation of pituitary function. As part of this evaluation, an x-ray examination was planned because of the possibility that a tumor was the cause of his excessive height of 7 feet as well as his abnormal weakness. The tests revealed that a pituitary tumor had resulted in the production of excess _____

2. Seventeen-year-old Ms. K had never had a menstrual period. The cause could have been a deficiency of the ovarian hormones called _____

3. Mrs. C, age 56, had been brought to the hospital in coma, that is, she was unconscious and could not be aroused. Tests revealed that her blood sugar was abnormally high. Mrs. C's illness was due to a lack of insulin, and is known as _____

4. Mrs. K consulted her doctor because she felt weak at times, especially after not eating for a while. She had also noted a darkening of the pigment in her skin. Tests showed low blood pressure and a deficiency of the hormones cortisol and aldosterone. These hormones are produced by the _____

11

The Blood

I. Overview

The blood maintains the internal environment in a constant state through its functions of transportation, regulation, and protection. Blood is composed of two elements; one, the liquid element or **plasma**, and the other, the **formed elements** consisting of the cells and cellular products. The plasma is 90% water and 10% proteins, carbohydrates, lipids, and mineral salts. The formed elements are composed of the **erythrocytes**, which carry oxygen to the tissues by means of their hemoglobin; the **leukocytes**, which defend the body against invaders; and the **platelets**, which are involved in the process of blood coagulation or clotting.

The forerunners of the blood cells are called **stem cells**. These are formed in the red bone marrow where they then develop into the various types of blood cells.

Blood **coagulation** is a protective mechanism that prevents blood loss when a blood vessel is ruptured by an injury. The first steps in the prevention of blood loss (hemostasis) include constriction of the blood vessels and formation of a platelet plug.

Should the quantity of blood in the body be severely reduced because of hemorrhage or disease, the cells suffer from lack of nourishment. In such instances, a **transfusion** may be given after typing and matching the blood of the recipient and donor. (Red cells with different surface proteins (**antigens**) than the recipient's red cells will react with **antibodies** in the recipient's blood, causing harmful agglutination and destruction of the donated red cells.) Blood can be packaged and stored in blood banks for use when transfusions are needed. Whenever possible, **blood components** such as cells, plasma, plasma fractions, or platelets are used. This practice is more efficient and reduces the chances of incompatibility.

The presence or absence of **Rh factor**, a red blood cell protein, is also important in transfusions. If blood containing the Rh factor (Rh positive) is given to a person whose blood lacks that factor (Rh negative), the recipient may become sensitized to the protein; his blood cells will produce antibodies to counteract the foreign substance. If an Rh negative mother becomes sensitized by an Rh positive fetus, her antibodies may damage the red cells of the fetus in a later pregnancy, resulting in **hemolytic disease of the newborn** (erythroblastosis fetalis).

Numerous **blood studies** have been devised in order to measure the composition of blood. These include the hematocrit, tests for the amount of hemoglobin, cell counts, and coagulation studies. Modern laboratories are equipped with automatic counters, which rapidly and accurately count blood cells, and with automatic analyzers, which measure enzymes, electrolytes, and other constituents of blood serum.

II. Topics for Review

A. Purposes of blood
B. Blood plasma and its functions
C. The formed elements and their functions
 1. Erythrocytes
  a. Structure
  b. Function
 2. Leukocytes
  a. Types
  b. Functions
 3. Platelets (thrombocytes)
D. Origin of blood cells
E. Blood clotting and hemostasis
F. Blood typing and blood transfusions
 1. Blood groups
 2. Rh factor
 3. Blood banks
 4. Blood components
G. Blood studies

III. Matching Exercises

Matching only within each group, print the answers in the spaces provided.

Group A

bone marrow	thrombocytes	carbon dioxide
oxygen	plasma	erythrocytes
hemoglobin	leukocytes	nucleus

1. The liquid part of the blood is known as　　　　　　　　　　

2. The red blood cells are called　　　　　　　　　　

3. There are several types of white blood cells or　　　　　　　　　　

4. Elements that have to do with clotting include platelets, or　　　　　　　　　　

5. An important gas that is transported by the blood from the lungs to all parts of the body is　　　　　　　　　　

6. The gaseous waste product carried by the blood to the lungs is the gas named　　　　　　　　　　

7. The iron-containing protein in red blood cells is a compound called

8. A connective tissue present in bone is the site of formation of blood cells. The name of this tissue is   _____

9. The mature red blood cell differs from other body cells in that it lacks a(n)   _____

Group B

albumin   cryoprecipitate   antigens
hemostasis   gamma globulin   Rh
neutrophils   plasmapheresis

1. The procedure for removing plasma and returning formed elements to a donor is   _____

2. An individual with hemophilia (a bleeding disease) may receive clotting factors in the form of a fraction obtained from frozen plasma. This fraction is called   _____

3. The most abundant protein in the blood is   _____

4. The disease erythroblastosis fetalis indicates an incompatibility between a mother and a fetus in the blood factor known as   _____

5. The most numerous leukocytes in the blood are   _____

6. A person who receives blood of a different type than his or her own may have antibodies to proteins on the surface of the red blood cells received. These proteins as a group are called   _____

7. Blood clotting is a step in the prevention of blood loss, a process called   _____

8. The fraction of the blood that contains antibodies is the fraction called   _____

Group C

type O   hemoglobin   pathogens
megakaryocytes   type AB   agglutination
fibrinogen   hemolysis

1. Oxygen, needed by all the tissues, is transported by a substance in red blood cells. This substance is called   _____

2. The platelets are fragments of large cells known as   _____

3. As platelets disintegrate they release a chemical that activates a plasma protein called   _____

4. The process whereby cells become clumped is known as   _____

5. In blood transfusion a dangerous condition that occurs when donor cells are dissolved or go into solution is   _____

6. Blood that is not clumped by either anti-A or anti-B serum belongs to the group called _____

7. If the cells are clumped by both the anti-A and anti-B serums the blood belongs to _____

8. The appearance of pus at a body site indicates that the leukocytes are actively involved in the destruction of _____

Group D

hyperglycemia	hemocytometer	transfusion
5,000 to 10,000	hematocrit	hemorrhage
centrifuge	4.5 to 5.5 million	blood chemistry

1. Another term for profuse abnormal bleeding is _____

2. The transfer of whole blood from one person to another is called a(n) _____

3. Separation of blood plasma from the formed elements of blood is accomplished by use of the _____

4. An apparatus made of several parts and used for counting blood cells is called a(n) _____

5. Normally, the number of red blood cells per cubic millimeter is _____

6. Normally, the number of white blood cells per cubic millimeter is _____

7. Blood normally contains some sugar. When the amount is excessive, the condition is referred to as _____

8. The volume percentage of red blood cells in centrifuged whole blood is called the _____

9. Measurements of electrolytes, blood urea nitrogen, enzymes, and glucose are included as part of _____

IV. Labeling

For each of the following illustrations, print the name or names of each labeled part on the numbered lines.

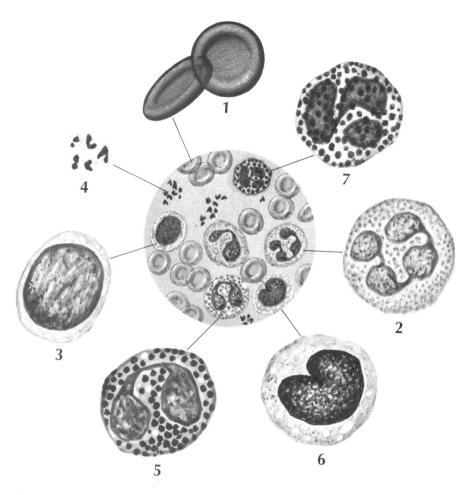

Blood cells

1. _____
2. _____
3. _____
4. _____
5. _____
6. _____
7. _____

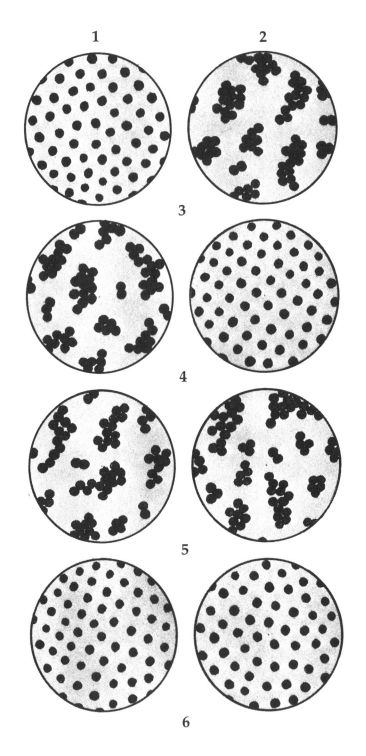

Blood typing

1. _____ 4. _____

2. _____ 5. _____

3. _____ 6. _____

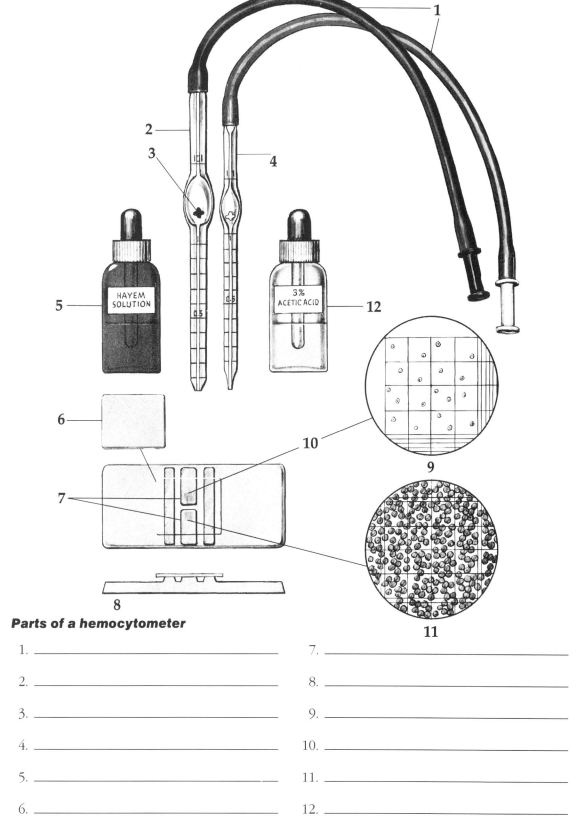

Parts of a hemocytometer

1. _____
2. _____
3. _____
4. _____
5. _____
6. _____
7. _____
8. _____
9. _____
10. _____
11. _____
12. _____

V. Completion Exercise

Print the word or phrase that correctly completes each sentence.

1. The gas that is transported to all parts of the body by the blood and that is necessary for life is called _____

2. One waste product of body metabolism is carried to the lungs to be exhaled. This gas is known as _____

3. Red blood cells are far more numerous than white ones; the proportion is about _____

4. A collection of dead and living white blood cells and bacteria in a region of infection is _____

5. Blood cells are formed in the _____

6. The number of different types of white blood cells is _____

7. Lymphocytes mature and may divide in the thymus and the _____

8. The process whereby cells are clumped together because of an incompatibility between red blood cells and another person's serum is called _____

9. A certain red blood cell protein is present in about 85% of the population. Such individuals are said to be _____

10. One of the transport functions of the blood is the transmission of a by-product of muscle activity from the muscles to all parts of the body. This by-product is _____

11. The most important function of certain lymphocytes is to engulf disease-producing organisms by the process of _____

12. Digested food proteins are absorbed into the capillaries of the intestinal villi in the form of protein building blocks, or _____

VI. Practical Applications

Study each discussion. Then print the appropriate word or phrase in the space provided.

Group A

1. Ms. G sustained numerous deep gashes when she accidentally broke a glass shower door. One of the cuts bled copiously. In describing this type of bleeding the doctor used the word _____

2. While the physician attended to the wound the technician drew blood for typing and other studies. Ms. G's blood was found to agglutinate with both anti-A and anti-B serum. Her blood was classified as group _____

3. Among the available donors were some whose blood was found to be free of both A and B surface antigens. They were classified as having blood type _____

4. Further testing of Ms. G's blood revealed that it lacked the Rh factor. She was therefore said to be _____

5. If Ms. G were to be given a transfusion of Rh positive blood, she might become sensitized to the Rh protein. In that event her blood would produce counteracting substances called _____

6. Mr. R had lost a large quantity of blood when he was injured in an automobile accident. In addition to whole blood, he was given several units of cryoprecipitate to replace lost _____

Group B

On the medical ward there were a number of patients who required extensive blood studies.

1. A boy 7 years of age had a history of frequent fevers and a tendency to bleed easily. Physical examination revealed enlarged lymph nodes. A blood smear revealed pronounced cell changes. The number of each kind of white cell was determined by counting the white cells in a sample of blood. This is called a _____

2. Mrs. C's history included rapid weight loss, constant thirst, and episodes of fainting. A blood test showed the presence of excessive sugar, or glucose. This condition is named _____

3. Mr. B, age 28, had a history of heart disease due to bacteria that caused dissolution (dissolving) of red blood cells. This type of disintegration is known as _____

4. Mr. K suffered from a viral infection of the liver. As a protective measure, his young son was given an injection of a protein substance obtained from human plasma. This antibody, which prevents certain viral infections, has the name of _____

12

The Heart

I. Overview

The ceaseless beat of the heart day and night throughout one's entire lifetime is such an obvious key to the presence of life that it is no surprise that this organ has been the subject of wonderment and poetry. When the heart stops pumping, life ceases. The cells must have oxygen, and it is the heart's pumping action that propels oxygenated blood to them.

In size the heart has been compared to a **closed fist**. In location it is thought of as being on the left side, although about one third is to the right of the midline. The muscular apex of the triangular heart is definitely on the left. It rests on the **diaphragm**, the dome-shaped muscle that separates the thoracic cavity from the abdominal space.

In birds and mammals, including humans, the heart has two sides, in which the aerated (higher in oxygen) and the unaerated (lower in oxygen) blood are kept entirely separated. So the heart is really a **double pump** in which the two sides pump in unison. The right side pumps blood to the lungs to be oxygenated, and the left side pumps blood to all other parts of the body.

Each side of the heart is divided into two parts or **chambers** which are in direct communication. The upper chamber or **atrium** on each side opens directly into the lower chamber or **ventricle**. **Valves** between the chambers keep the blood flowing forward as the heart pumps. The atria are the receiving chambers for blood returning to the heart. The two ventricles pump blood to all parts of the body. Because they pump more forcefully, their walls are thicker than the walls of the atria. The coronary arteries supply blood to the heart muscle or **myocardium**.

The heartbeat originates within the heart at the **sinoatrial node**, often called the pacemaker. Electrical impulses from the pacemaker spread over special fibers in the wall of the heart to produce contractions, first of the two atria and then of the two ventricles. After contraction, the heart relaxes and fills with blood. The relaxation phase is called **diastole** and the contraction phase is called **systole**. Together these two phases make up one **cardiac cycle**.

II. Topics for Review

A. The heart as a pump
B. Structure of the heart wall
    1. Endocardium
    2. Myocardium
    3. Epicardium
C. The pericardium
D. Anatomy of the heart
    1. Septum
    2. Chambers
    3. Valves
E. The cardiac cycle
    1. Diastole
    2. Systole
F. The conduction system of the heart
    1. Sinoatrial node (pacemaker)
    2. Atrioventricular node
    3. Atrioventricular bundle and bundle branches
G. Normal and abnormal heart sounds
H. Instruments used to study the heart

III. Matching Exercises

Matching only within each group, print the answers in the spaces provided.

Group A

tricuspid valve	interatrial septum	aortic valve
endocardium	myocardium	interventricular septum
mitral valve	epicardium	pulmonary semilunar valve

1. The membrane of which the heart valves are formed and which lines the interior of the heart is called _____

2. By far the thickest layer in the heart wall is the muscular one, the _____

3. The outermost layer of the heart is the _____

4. A partition, the septum, separates the two sides of the heart. The thin-walled upper part of this septum is the _____

5. The larger part of the partition between the two sides of the heart is the _____

6. Between the two right chambers of the heart lies the right atrioventricular valve. It is also called the _____

7. The left atrioventricular valve is thicker and heavier than the right; it is made of two flaps or cusps. It is called the _____

8. Situated between the right ventricle and the pulmonary artery is the valve that prevents blood on its way to the lungs from returning to the right ventricle. This is the _____

9. The valve that prevents blood from returning after the left ventricle has emptied itself is the _____

Group B

arteries   sinoatrial node   systole
atria   veins   diastole
atrioventricular node   atrioventricular bundle

1. The contraction phase of the cardiac cycle is called _____

2. The brief resting period that follows the contraction phase of the cardiac cycle is _____

3. Impulses in the heart follow a definite sequence, beginning in the pacemaker. The pacemaker is located in the upper right atrial wall and is called the _____

4. Next, the excitation wave travels throughout the muscles of the upper heart chambers, causing them to contract. These are the _____

5. Following this, the second mass of conduction tissue (located in the septum) is stimulated. This is the _____

6. Finally, the ventricular musculature contracts in response to stimulation by branches of the _____

7. Blood is pumped to the lungs and body tissues through _____

8. Oxygenated blood from the lungs and deoxygenated blood from the body tissues is carried through the _____

Group C

stroke volume   tachycardia   ventricles
murmur   bradycardia   cardiac output
myocardium

1. A heart rate of greater than 100 beats per minute is described as _____

2. The pumping chambers of the heart are the _____

3. The coronary arteries supply blood to the _____

4. The volume of blood pumped by each ventricle in 1 minute is the _____

5. A heart rate of less than 60 beats per minute is called _____

6. Abnormal closing of the heart valves may result in a(n) _____

7. The amount of blood ejected from a ventricle with each beat is the _____

Group D

fluoroscope septum echocardiography
functional stethoscope electrocardiograph

1. The type of murmur that is not associated with abnormalities of the heart is described as _____

2. An instrument for recording the electrical activity of the heart is the _____

3. An instrument that uses x-rays in examining deep structures is the _____

4. The simple instrument used by the physician for listening to sounds from within the patient's body is the _____

5. The partition between the two sides of the heart is the _____

6. A rapid, painless, and harmless method for studying the heart uses sound impulses that are reflected and recorded. This is _____

IV. Labeling

For each of the following illustrations print the name or names of each labeled part on the numbered lines.

1. _____   7. _____

2. _____   8. _____

3. _____   9. _____

4. _____   10. _____

5. _____   11. _____

6. _____   12. _____

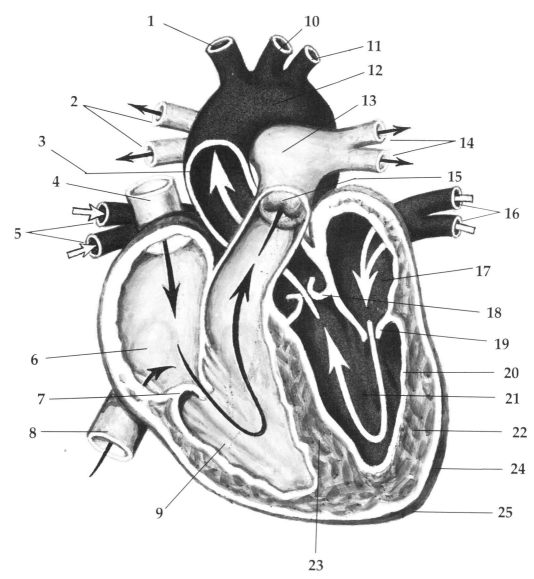

The heart and great vessels

13. _____
14. _____
15. _____
16. _____
17. _____
18. _____
19. _____

20. _____
21. _____
22. _____
23. _____
24. _____
25. _____

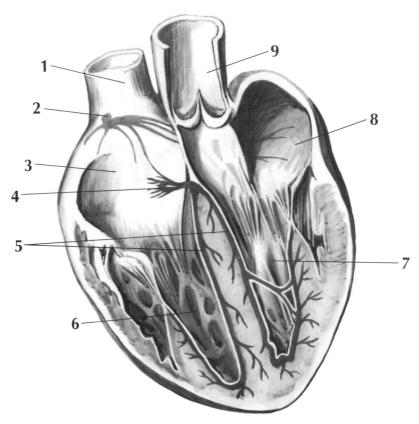

Conduction system of the heart

1. _____
2. _____
3. _____
4. _____
5. _____
6. _____
7. _____
8. _____
9. _____

V. Completion Exercise

Print the word or phrase that correctly completes each sentence.

1. The continuous one-way movement of the blood is known as the _____

2. Each minute the heart contracts on an average of about _____

3. The fibrous sac that surrounds the heart is the _____

4. After supplying nutrients to the heart muscle, blood is drained into the right atrium by way of the _____

5. The partition between the two thick-walled lower chambers of the heart is the _____

6. Because each of the three parts of the two exit valves is half-moon shaped, these valves are described as _____

7. The right atrioventricular valve is the _____

8. One complete sequence of relaxation and contraction of the heart is called a(n) _____

9. The stroke volume and heart rate determine the volume of blood pumped by each ventricle in 1 minute, the _____

10. The main influence over the heart rate outside of the heart itself is the _____

VI. Practical Applications

Study each discussion. Then print the appropriate word or phrase in the space provided.

1. Mrs. K had suffered several attacks of rheumatic fever. The examining physician found evidence of damage to the membranous tissue of the valves. This membrane is called _____

2. Using a stethoscope, the physician listened for the heart sounds. The first sound occurs during the contraction phase of the cardiac cycle. This active phase is called _____

3. The second sound includes the effects of valve closure and occurs during the resting part of the cycle. This phase is called _____

4. Mr. L, age 62, complained of weakness and fatigue. Tests revealed damage to the conduction system of the heart. Normally, the beginning of the heartbeat occurs in the natural pacemaker, which is also called the _____

5. The greatest damage was found in the bundle of His, which is also known as the _____

6. One of the first tests that was done on both these patients was a recording of electrical currents produced by heart muscle. The apparatus that records this information is the _____

13

Blood Vessels and Blood Circulation

I. Overview

The blood vessels are classified, according to function, as **arteries**, **veins**, or **capillaries**; the arteries and veins are subdivided into pulmonary vessels and systemic vessels.

The two arterial systems, the **systemic** and the **pulmonary**, can be likened to trees: each has a trunk, the aorta in one and the pulmonary artery in the other. Each trunk has subdivisions, large and small branches that carry the blood into the capillaries where exchanges between the blood and the tissue fluid occur. The tissue fluid provides for the transfer of substances required by the cell in exchange for those not needed or those manufactured for use elsewhere. The venous systems consist of tributaries progressing in size from small to large; they return the blood to the heart, which pumps it into the arterial trunks, thus completing the circuit.

The walls of the vessels, especially the small arteries, contain smooth muscle which is under the control of the involuntary nervous system. The diameters of the vessels can be regulated by the nervous system to alter blood pressure and to direct blood to various parts of the body as needed. These changes, termed **vasodilation** and **vasoconstriction**, are centrally controlled by a **vasomotor center** in the medulla of the brainstem.

The walls of the arteries are thicker and more elastic than the walls of the veins, and the arteries contain blood under higher pressure. All vessels are lined with a single layer of simple epithelium called **endothelium**. The smallest vessels, the capillaries, are made only of this single layer of cells. It is through the walls of the capillaries that exchanges take place between the blood and the tissues.

The **pulse rate** and **blood pressure** are manifestations of the circulation; they tell the trained person a great deal about the overall condition of the individual.

II. Topics for Review

A. The blood vessels
   1. Structure and function
B. Pulmonary and systemic circuits

C. Systemic arteries
   1. Branches of the aorta
      a. Ascending
      b. Aortic arch
      c. Thoracic
      d. Abdominal
   2. Branches of the iliac arteries
   3. Other parts of the arterial tree
D. Anastomoses
E. Systemic veins
   1. Superficial
   2. Deep
   3. Superior vena cava
   4. Sinuses
   5. Inferior vena cava
F. The hepatic portal system
G. Capillary exchanges
H. Vasodilation and vasoconstriction
I. Return of blood to the heart
J. Pulse
K. Blood pressure

III. Matching Exercises

Matching only within each group, print the answers in the spaces provided.

Group A

systemic	endothelium	pulmonary
arteries	celiac	carotid
aorta	capillaries	coronary

1. The vessels that are related to the lungs, including the arteries and their branches in the lungs and the veins that drain lung capillaries are all designated as _____

2. Exchanges between the blood and the cells take place through the _____

3. Since their function is to carry blood from the heart's pumping chambers, the blood vessels that have the thickest walls are the _____

4. The innermost tunic of the artery is composed of _____

5. The largest artery in the body is divided into four regions. This vessel is the _____

6. The arteries that carry food and oxygen to the tissues of the body are classified as _____

7. The ascending aorta has two branches that supply the heart muscle; they are described by the term _____

8. Supplying the head and neck on each side is an artery named the _____

9. One of the unpaired arteries that supplies some of the viscera of the upper abdomen is a short trunk, the _____

Group B

blood pressure	portal	anastomosis
valves	arterioles	vasomotor center
venules		

1. Small arteries are called _____

2. The vessels that receive blood from the capillaries are the _____

3. A system that carries venous blood to a second capillary bed before it returns to the heart is described by the term _____

4. Dilation and constriction of the blood vessels is controlled by an area of the medulla called the _____

5. A communication between two arteries is a(n) _____

6. A force that drives materials out of the capillaries is _____

7. Blood is prevented from moving backward in the veins by the _____

Group C

phrenic artery	lumbar arteries	common iliac arteries
brachiocephalic trunk	right subclavian artery	left common carotid artery
renal arteries	hepatic artery	superior mesenteric artery
brachial artery		

1. Coming off the aortic arch is a short artery formerly called the innominate artery. This is the _____

2. Supplying the left side of the head and neck is the _____

3. Oxygenated blood is carried to the liver by the _____

4. The largest branch of the abdominal aorta supplies most of the small intestine and the first half of the large intestine. This branch is the _____

5. The diaphragm is supplied by a right and a left _____

6. The artery supplying the arm is a continuation of the axillary artery, and is called the _____

7. Blood supply to the right upper extremity is through the _____

8. The largest of the paired branches of the abdominal aorta are those that supply the kidneys. These are the _____

9. The abdominal aorta finally divides into two _____

10. Supply to the abdominal wall is through the _____

Group D

brachiocephalic trunk
mesenteric
paired
celiac trunk
radial artery
femoral artery
circle of Willis
unpaired
basilar artery
volar arch

1. An anastomosis of the two internal carotid arteries and the basilar artery is located immediately under the center of the brain. It is called the _____

2. The inferior mesenteric is an example of an artery that is _____

3. The radial and ulnar arteries in the hand anastomose to form the _____

4. Anastomoses between branches of the vessels supplying blood to the intestinal tract comprise arches named _____

5. The right subclavian artery and the right common carotid artery are branches of the _____

6. The left gastric artery and the splenic artery are two of the three branches of the _____

7. The union of the two vertebral arteries forms the _____

8. The external iliac arteries extend into the thigh. Here each of them becomes a(n) _____

9. The popliteal arteries are examples of the many blood vessels that are _____

10. The branch of the brachial artery that extends down the forearm and wrist of the thumb side is the _____

Group E

azygos vein
inferior vena cava
jugular veins
median cubital
hepatic portal vein
brachiocephalic veins
saphenous vein
superior vena cava
venous sinuses

1. The longest vein is the superficial one called the _____

2. Because of its location near the surface at the front of the elbow, one of the veins frequently used for removing blood for testing is the _____

3. The areas supplied by the carotid arteries are drained by the _____

4. The union of the jugular and subclavian veins forms the _____

5. Veins draining the head, the neck, the upper extremities, and the chest all empty into the _____

6. Before reaching the superior vena cava (and then the heart), blood from the chest wall drains into the _____

7. Large channels that drain deoxygenated blood are called _____

8. The blood from the parts of the body below the diaphragm is drained by the large vein called the _____

9. Tributaries from the unpaired organs empty into a vein that enters the liver where it subdivides into smaller veins. This unusual vein is called the _____

Group F

sinusoids
coronary sinus
common iliac veins
superior mesenteric vein
hepatic veins
lateral sinuses
superior sagittal sinus
left testicular vein
cavernous sinuses
gastric veins

1. The inferior vena cava begins with the union of the two _____

2. The only exceptions to the rule that paired veins empty directly into the vena cava are the left ovarian vein and the _____

3. Among the paired veins that empty directly into the inferior vena cava are those draining the liver, the _____

4. The vein that drains most of the small intestine and the first part of the large intestine is the _____

5. The tributaries of the hepatic portal vein include those that drain the stomach, the _____

6. Within the liver, exchanges take place through enlarged capillaries called _____

7. The veins of the heart drain mainly into the _____

8. The ophthalmic veins drain into the _____

9. Nearly all the blood from the veins of the brain eventually empties into one or the other of the transverse or _____

10. In the midline above the brain and in the fissure between the two cerebral hemispheres is a long blood-containing space called the _____

Group G

systolic pressure   faster   sphygmomanometer
slower   diastolic pressure pulse
dorsalis pedis   radial artery

1. Beginning at the heart and traveling along the arteries is a wave of increased pressure started by the force of ventricular contractions. This wave is called the _____

2. The wave is readily felt at the wrist because of the artery that passes over the bone on the thumb side. This is the _____

3. Sometimes it is necessary to use the artery on the top of the foot for obtaining the pulse. This is the _____

4. Blood pressure is recorded by an instrument called a(n) _____

5. It is important to recognize factors that may influence pulse rate. Emotional disturbance, for example, may cause the pulse rate to be _____

6. As the child matures, his pulse rate normally becomes _____

7. During ventricular relaxation the sphygmomanometer measures _____

8. Blood pressure measured during heart muscle contraction is the _____

IV. Labeling

For each of the following illustrations print the name or names of each labeled part on the numbered lines.

1. _____   11. _____
2. _____   12. _____
3. _____   13. _____
4. _____   14. _____
5. _____   15. _____
6. _____   16. _____
7. _____   17. _____
8. _____   18. _____
9. _____   19. _____
10. _____   20. _____

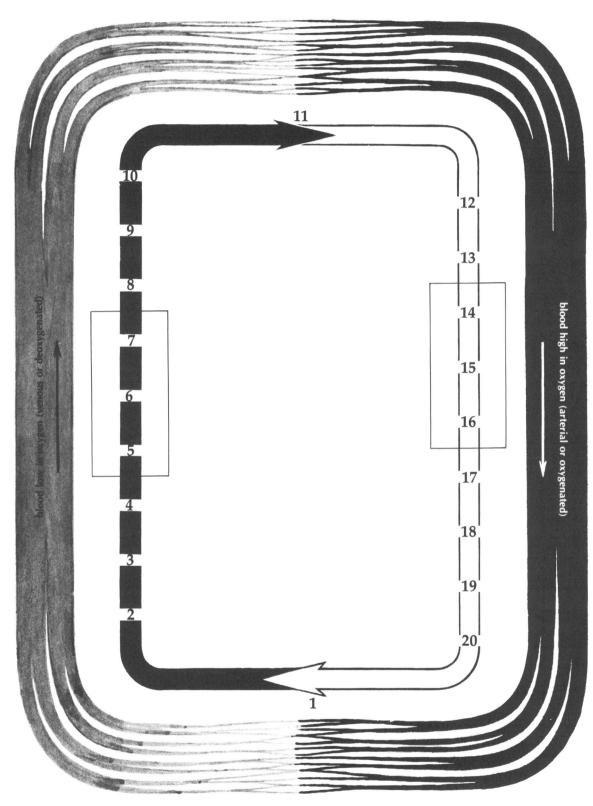

Diagram to show circuit of blood flow

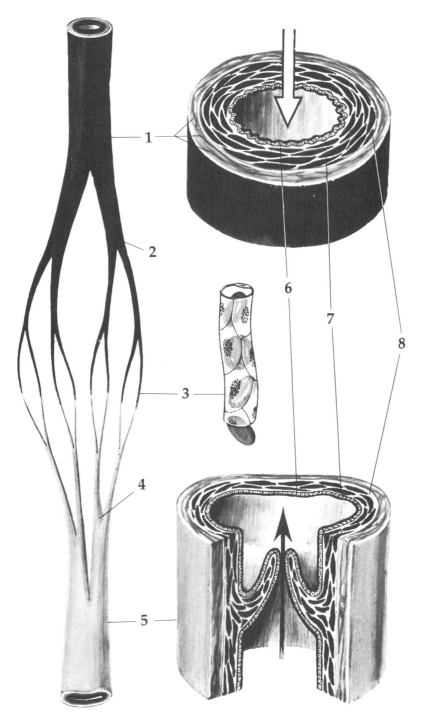

Sections of small blood vessels

1. _____
2. _____
3. _____
4. _____
5. _____
6. _____
7. _____
8. _____

1. _____
2. _____
3. _____
4. _____
5. _____
6. _____
7. _____
8. _____
9. _____
10. _____
11. _____
12. _____
13. _____
14. _____
15. _____
16. _____
17. _____
18. _____
19. _____
20. _____
21. _____

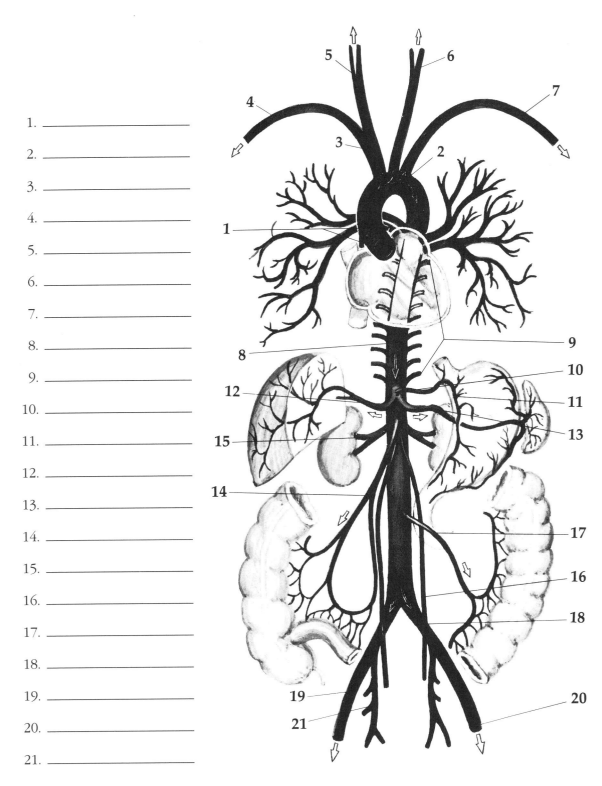

The aorta and its branches

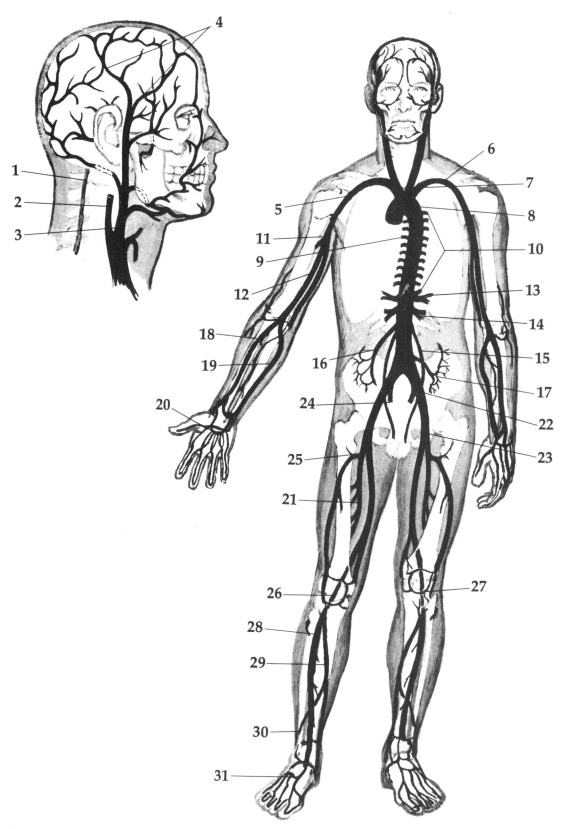

Principal arteries

1. _____   17. _____
2. _____   18. _____
3. _____   19. _____
4. _____   20. _____
5. _____   21. _____
6. _____   22. _____
7. _____   23. _____
8. _____   24. _____
9. _____   25. _____
10. _____   26. _____
11. _____   27. _____
12. _____   28. _____
13. _____   29. _____
14. _____   30. _____
15. _____   31. _____
16. _____

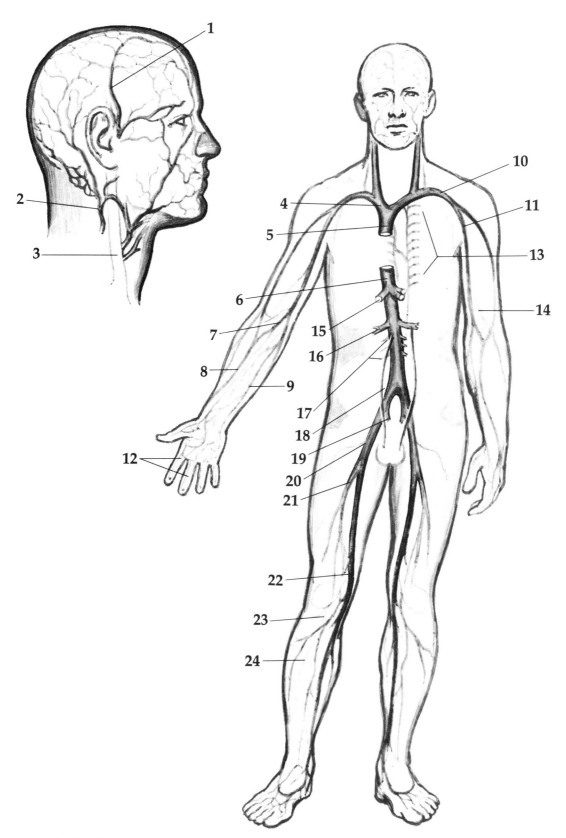

Principal veins

1. _____	13. _____
2. _____	14. _____
3. _____	15. _____
4. _____	16. _____
5. _____	17. _____
6. _____	18. _____
7. _____	19. _____
8. _____	20. _____
9. _____	21. _____
10. _____	22. _____
11. _____	23. _____
12. _____	24. _____

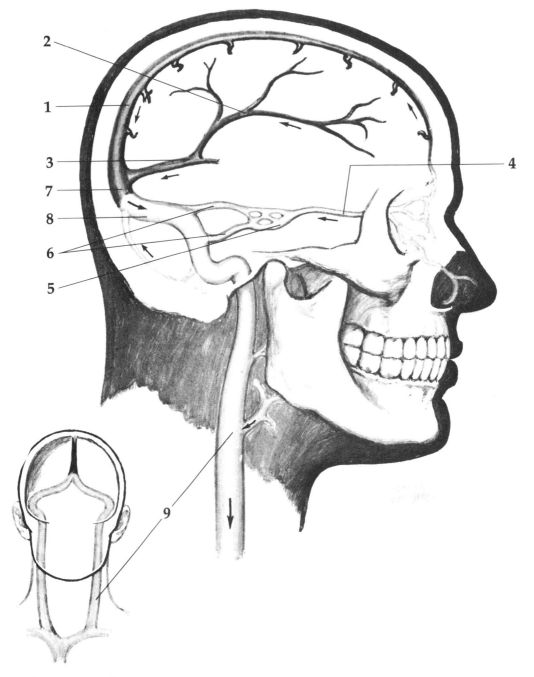

Cranial venous sinuses

1. _____

2. _____

3. _____

4. _____

5. _____

6. _____

7. _____

8. _____

9. _____

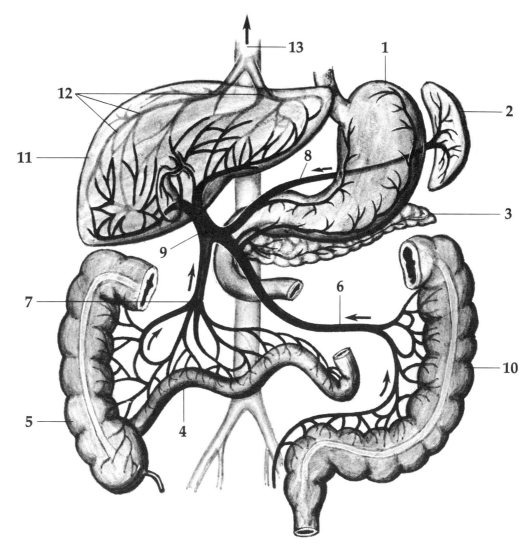

Hepatic portal circulation

1. _____
2. _____
3. _____
4. _____
5. _____
6. _____
7. _____
8. _____
9. _____
10. _____
11. _____
12. _____
13. _____

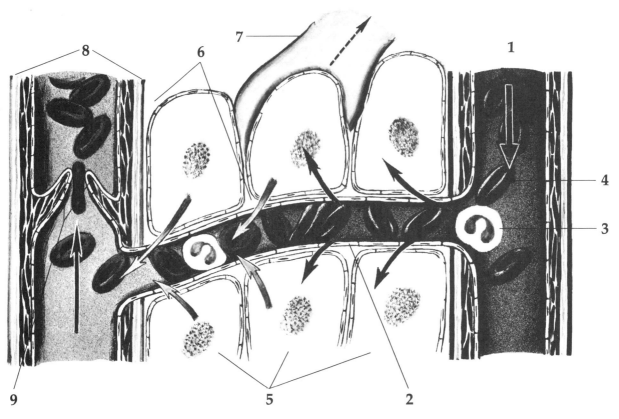

Diagram showing the connection between the small blood vessels through capillaries

1. _____
2. _____
3. _____
4. _____
5. _____
6. _____
7. _____
8. _____
9. _____

V. Completion Exercise

Print the word or phrase that correctly completes each sentence.

1. Deoxygenated blood is carried from the right ventricle by the _____

2. The smallest subdivisions of arteries have thin walls in which there is little connective tissue and relatively more muscle. These vessels are _____

3. Supplying nutrients to body tissues and carrying off waste products from the tissues are functions of the part of the circulation described as _____

4. The middle tunic of the arterial wall is composed of elastic connective tissue plus _____

5. The innermost tunic of the vessels is composed of _____

6. The smallest veins are formed by the union of capillaries. These tiny vessels are called _____

7. The circle of Willis is formed by a union of the internal carotid arteries and the basilar artery. Such a union of end arteries is called a(n) _____

8. An increase in the diameter of a blood vessel is termed _____

9. One example of a portal system is the system which carries blood from the abdominal organs to the _____

VI. Practical Applications

Study each discussion. Then print the appropriate word or phrase in the space provided.

1. Mr. S, age 53, complained of shortness of breath, weakness, and pain in the left chest. Examination indicated that the left semilunar valve was not functioning properly. This valve guards the entrance into the largest artery which is the _____

2. Mr. K, age 71, was admitted to the hospital because of fainting attacks and inability to recall events just before and after these episodes. Physical examination revealed some interference with the normal flow through the large arteries that carry blood to the brain. These neck arteries are named _____

3. Small amounts of blood could still reach Mr. K's brain through the two vertebral arteries. These join at the base of the brain to form a single artery called the _____

4. Mr. K was protected from brain death by the fact that three arteries join to form an anastomosis under the brain. This anastomosis is the _____

5. Mr. J, an alcoholic, came in complaining of a variety of digestive and nervous disturbances. Examination revealed an enlarged liver and an accumulation of fluid in the abdominal cavity. There was evidence of obstruction of the large vein that drains the unpaired organs of the abdomen. This large vessel is called the _____

6. Mr. J and his digestive problems were studied for evidence of back pressure within the veins that drain the intestine. The larger of these is called the _____

7. Mrs. B, age 67, had been diabetic for several years. Now there was evidence of artery disease in several of the larger vessels. One of these is named according to regions including the portion nearest the heart, called the _____

8. Another vessel that was seriously damaged by hardening was the short one that is the first branch of the aortic arch. It is the _____

14

The Lymphatic System and Immunity

I. Overview

Lymph is the watery fluid that flows within the lymphatic system. It originates from the blood plasma and from the tissue fluid that is found in the minute spaces around and between the body cells. The fluid moves from the **lymphatic capillaries** through the **lymphatic vessels** and then to the **right lymphatic duct** and the **thoracic duct**. The lymphatic vessels are thin-walled and delicate; like some veins, they have valves that prevent backflow of lymph.

The **lymph nodes**, which are the system's filters, are composed of **lymphoid tissue**. These nodes remove impurities and manufacutre **lymphocytes**, cells active in immunity. Chief among them are the cervical nodes in the neck, the axillary nodes in the armpit, the tracheobronchial nodes near the trachea and bronchial tubes, the mesenteric nodes between the peritoneal layers, and the inguinal nodes in the groin area.

In additon to these nodes, there are several organs of lymphoid tissue with somewhat different functions. The **tonsils** filter tissue fluid. The **thymus** is essential for development of the immune system during the early months of life. The **spleen** has numerous functions, among which are the destruction of worn-out red blood cells, serving as a reservoir for blood, and the production of red blood cells before birth.

Another part of the body's protective system is the **mononuclear phagocyte system**, which consists of cells involved in the destruction of bacteria, cancer cells, and other possibly harmful substances.

The ultimate defense against disease is **immunity**, the means by which the body resists or overcomes the effects of a particular disease or other harmful agent. Immune responses are based on reactions between foreign substances or **antigens** and certain white blood cells, the lymphocytes. The **T-lymphocytes** respond to the antigen directly and produce **cell-mediated immunity**. **B-lymphocytes**, when stimulated by an antigen, multiply and begin to produce specific antibodies which react with the antigen. These circulating antibodies make up the form of immunity termed **humoral immunity**.

There are two basic types of immunity: inborn and acquired. **Inborn immunity** is inherited; it may exist on the basis of **species**, **racial**, or **individual** characteristics. Acquired immunity may be of the **natural** type (acquired before birth or by contact with the disease) or of the **artificial** type (provided by a vaccine or an immune serum). Immunity acquired as a result of the transfer of antibodies to an individual from some outside source is described as **passive immunity**; immunity that involves production of antibodies by the individual is termed **active immunity**.

II. Topics for Review

A. Lymphatic vessels
B. Lymphoid tissue
   1. Lymph nodes
   2. Tonsils
   3. Thymus
   4. Spleen
C. Functions of the mononuclear phagocyte system
   1. Other nonspecific defenses
D. Immunity
   1. Inborn
   2. Acquired
      a. Active
      b. Passive
E. Lymphocytes and the immune response
F. Vaccines and immunization

III. Matching Exercises

Matching only within each group, print the answers in the spaces provided.

Group A

right lymphatic duct   cervical nodes   inguinal nodes
lacteals   blood   chyle
endothelium   valves   axillary nodes

1. There is easy passage of soluble materials and water through the walls of lymphatic capillaries, composed of a single layer of cells forming the _____

2. The lymphatics resemble some veins in that they contain structures that prevent backflow. These are _____

3. One pathway for fats from digested food to the bloodstream is through specialized lymphatic capillaries of the intestine that are called _____

4. Lymph is drained from the right side of the head, of the neck, of the thorax, and of the right upper extremity by the _____

5. The combination of fat globules and lymph gives rise to a milky-appearing fluid called _____

6. Lymph nodes are named according to location. Those located in the armpits are known as _____

7. Lymph from the lower extremities and the external genitalia drains through the _____

8. The final destination of filtered lymph is the _____

9. The lymph nodes located in the neck and draining certain parts of the head and neck are known as _____

Group B

thymus   pharyngeal tonsils   palatine tonsils
lingual tonsils   phagocytosis   antibodies
spleen   monocytes

1. The oval lymphoid bodies located at each side of the soft palate are known as _____

2. The enlarged masses of lymphoid tissue often found on the back wall of the pharynx and commonly called adenoids are correctly called _____

3. At the back of the tongue are masses of lymphoid tissue called _____

4. The structure that is believed to be essential in the development of immunity very early in life is the _____

5. Blood is filtered by an organ located in the upper left quadrant (left hypochondriac region) of the abdomen. This is the _____

6. The white blood cells active in the mononuclear phagocyte system are the _____

7. Some of the cells in the lymphoid tissue produce substances that aid in combating infection. These are called _____

8. The spleen and other lymphoid tissues generate cells that are able to engulf bacteria and other small foreign bodies. This process of ingesting foreign substances is called _____

Group C

lymph nodes   backflow   radial
subclavian vein   drainage   thymus
thoracic duct

1. T-lymphocytes mature in the _____

2. Before the lymph reaches the veins it is passed through organs that act as filters. These are _____

3. The lymphatic vessels serve as a system for _____

4. Lymph received in the right lymphatic duct drains into the right

5. Lymph is drained from the body below the diaphragm and on the left side above the diaphragm by the largest lymphatic vessel, the _____

6. The valves of the lymphatic vessels prevent lymph _____

7. Lymphatic vessels are named according to location; thus, those on the lateral side of the forearm are described as _____

Group D

lymphocytes   lacteals   lymph nodes
lymph   superficial   cisterna chyli
veins   plasma

1. Lymphatic vessels located under the skin are described as _____

2. The first part of the thoracic duct is enlarged, forming a temporary storage area. It is called the _____

3. Chyle, the fluid formed by combination of lymph and fat globules, comes from the intestinal _____

4. An important function of lymph nodes is the manufacture of white blood cells known as _____

5. Intercellular fluid originates from the liquid part of the blood. The liquid part of the blood is called _____

6. Tissue fluid passes from the intercellular spaces into the lymphatic vessels; it is then called _____

7. The masses of lymphoid tissue that filter foreign substances from the lymph are known as _____

8. The vessels of the lymphatic system are often located near the _____

Group E

macrophages   spleen   exercise
thoracic duct   chyle   Kupffer's cells
hilum

1. The area of exit for the vessels carrying lymph out of the node is known as the _____

2. When monocytes enter the body tissues and act to destroy foreign matter, they are called _____

3. The phagocytes of the liver are called _____

4. The normal onward flow of lymph is aided by changes in position of various parts of the body and by _____

5. A fluid formed by the combination of fat globules and lymph originates in the intestinal lacteals. The name of the fluid is _____

6. The larger of the terminal vessels of the lymphatic system is called the _____

7. During embryonic and fetal life, red blood cells are produced partly in the _____

Group F

| immunity | B-lymphocytes | nonspecific resistance |
| toxins | antigen | thymus |

1. The unbroken skin and mucous membranes are elements of the body's defenses against any harmful agent. Such protective devices are part of the body's _____

2. The poisons produced by pathogens are known as _____

3. The body's ability to defend itself against a certain specific agent is spoken of as its specific resistance, or _____

4. T cells are formed from stem cells that migrate from the bone marrow into the _____

5. Any foreign substance introduced into the body by a pathogen provokes a response. The foreign substance is known as a(n) _____

6. In response to infection, antibodies are produced by the _____

Group G

| inborn immunity | active immunity | passive immunity |
| species immunity | racial immunity | attenuation |

1. Animals are susceptible to diseases that do not affect humans; the reverse is also true. In other words, both possess a(n) _____

2. The greater resistance of black Americans to malaria and yellow fever over white Americans is an example of _____

3. An inherited immunity is usually called a(n) _____

4. A person who is infected by a pathogen or its toxin produces antibodies that make him immune to that infection for a long period of time, perhaps for life. This type of immunity is called _____

8. To produce immunity by vaccination, a live weakened pathogen may be administered. The process of reducing the virulence of the pathogen is called _____

6. The resistance of the newborn infant to contagious diseases is due to antibodies transferred from mother to fetus through the placenta. This type of immunity is classified as _____

Group H

complement memory cells toxoid
plasma cells vaccination gamma globulin
artificially acquired antigen

1. A person who has been bitten by an animal suspected of having rabies is immunized with a killed rabies virus vaccine. Such treatment is a form of _____

2. Some Rh negative persons may become sensitized to the Rh protein. In such cases the Rh factor is an example of a substance known as a(n) _____

3. The administration of a vaccine results in the type of immunity classified as _____

4. A toxin may be used for a vaccine if it is treated with heat or chemicals to reduce its harmfulness. Such an altered toxin is a(n) _____

5. The fraction of the blood plasma that contains antibodies is known as _____

6. When B-lymphocytes are stimulated by an antigen, they multiply and become cells capable of producing antibodies. These cells are called _____

7. Booster shots are given to stimulate the activity of _____

8. A group of blood proteins that may be needed to help an antibody destroy a foreign antigen is called _____

IV. Labeling

Print the name or names of each labeled part on the numbered lines.

1. _____ 10. _____

2. _____ 11. _____

3. _____ 12. _____

4. _____ 13. _____

5. _____ 14. _____

6. _____ 15. _____

7. _____ 16. _____

8. _____ 17. _____

9. _____

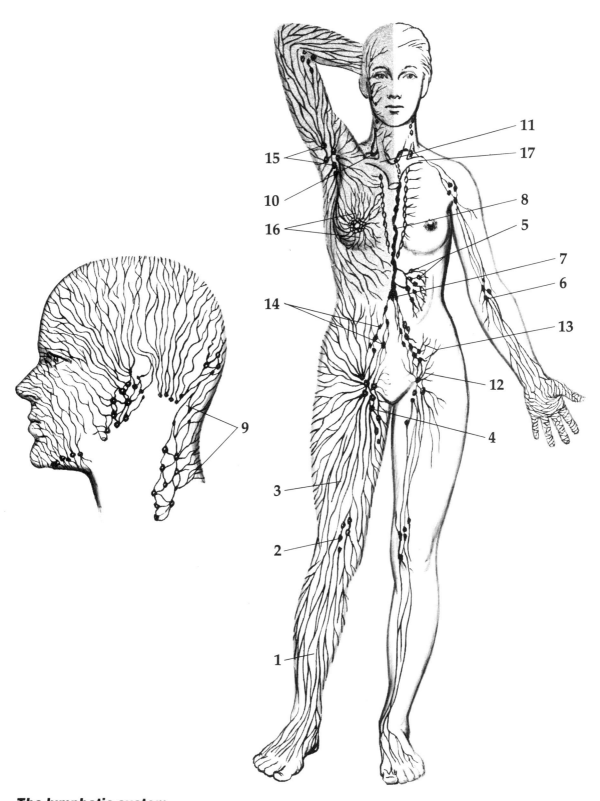

The lymphatic system

151

V. Completion Exercise
Print the word or phrase that correctly completes each sentence.

1. The fluid that moves from blood plasma to the tissue spaces and finally to special collection vessels is called _____

2. Lymphatic vessels from the left side of the head, neck, and thorax empty into the largest of the lymph vessels, the _____

3. The lymph from the body below the diaphragm and from the left side above the diaphragm is carried into the blood of the _____

4. Between the two layers of peritoneum that form the mesentery are found nodes called _____

5. In city dwellers, nodes may appear black because they become filled with carbon particles. This is true mostly of the nodes that surround the windpipe and its divisions. These are the _____

6. The structure popularly known as adenoids is correctly called the _____

7. The action of leukocytes in which they engulf and digest invading pathogens is known as _____

8. Circulating antibodies are responsible for the type of immunity termed _____

9. Immunity is a selective process through which a person may be immune to one disease but not to another. This selective characteristic is called _____

10. There are two main categories of immunity. One is inborn immunity while the other is _____

11. An artificially acquired immunity may be provided for a limited time by injecting an immune serum or for longer by using a _____

12. Antibodies transmitted from the mother's blood to the fetus provide a type of short-term borrowed immunity called _____

13. The administration of vaccine, on the other hand, stimulates the body to produce a longer lasting type of immunity called _____

15 Respiration

I. Overview

Oxygen is supplied to the tissue cells, and carbon dioxide is removed from them by means of the spaces and passageways that make up the **respiratory system**. This system contains the **nasal cavities**, the **pharynx**, the **larynx**, the **trachea**, the **bronchi**, and the **lungs**.

The two phases of breathing are the active phase of **inhalation** and the passive phase of **exhalation**. The movement of air between the outside and the lungs is called **ventilation**. As a result of ventilation, exchanges of oxygen and carbon dioxide occur. The first phase of these exchanges takes place in the lungs and is termed **external respiration**. The second phase, **internal respiration**, involves the exchanges that go on between the blood and the tissue cells.

The exchanges of oxygen and carbon dioxide are based on the physical principle of movement of molecules from areas of higher concentration to those of lower concentration by **diffusion**. Oxygen is transported to the tissues almost entirely by the **hemoglobin** in red blood cells. Some carbon dioxide is transported in the red blood cells as well, but most is carried in the blood plasma as the **bicarbonate ion**. Carbon dioxide is important in regulating the pH of the blood and in regulating the breathing rate.

Breathing is basically controlled by the **respiratory control centers** in the medulla and the pons of the brain stem. These centers are influenced by chemoreceptors located outside the medulla that respond to changes in the acidity of the cerebrospinal fluid. There are also **chemoreceptors** in the large vessels of the chest and neck that regulate respiration in response to changes in the composition of the blood.

II. Topics for Review

A. The respiratory system
B. Pulmonary ventilation
   1. Inhalation
   2. Exhalation

C. Gas exchanges
   1. External respiration
   2. Internal respiration
D. Gas transport
   1. Oxygen
   2. Carbon dioxide
E. Regulation of respiration
   1. Respiratory rates

III. Matching Exercises

Matching only within each group, print the answers in the spaces provided.

Group A

diffusion   inhalation   oxygen
exhalation   internal respiration   carbon dioxide
ventilation   external respiration   surfactant

1. The word *respiration* means "to breathe again"; one of its basic purposes is to supply the body cells with the gas   _____

2. At the same time that the required gas is being supplied, another gas, a waste product of cell metabolism, is being removed. This waste gas is   _____

3. The aspect of respiration involving gas exchanges in the lungs is called   _____

4. The second aspect of respiration refers to gas exchanges between the blood and body cells. This is called   _____

5. The movement of air into and out of the lungs is termed   _____

6. The physiology of respiration involves two phases of breathing. In the first phase air is drawn into the lungs. This is   _____

7. In the second phase of breathing air is expelled from the alveoli. This phase is called   _____

8. The substance in the fluid lining the alveoli that prevents their collapse is   _____

9. Gas exchange depends on the movement of gases from areas where they are in higher concentration to areas where they are in lower concentration. This physical change is named   _____

Group B

septum   larynx   trachea
pharynx   conchae   vascular
diaphragm   sinuses

1. Below the nasal cavities is an area that is common to both the digestive and respiratory systems. This is the _____

2. The cartilaginous structure commonly referred to as the voice box has the scientific name of _____

3. Several parts of the respiratory tract are kept open by a framework of cartilage. One of these is the windpipe or _____

4. The partition separating the two nasal cavities is called the nasal _____

5. The surface over which the air moves is increased by three projections located at the lateral walls of each nasal cavity. These are the _____

6. The lining of the nasal cavities contains many blood vessels and is therefore described as _____

7. The small cavities in the bones of the skull lined with mucous membrane are called _____

8. Separating the thoracic cavity from the abdominal cavity is the muscular _____

Group C

hilum (or hilus)   epiglottis   vocal cords
bronchi   esophagus   nasopharynx
ciliated epithelium   oropharynx   chemoreceptors

1. The mucous membranes lining the tubes of the respiratory system are mostly made of _____

2. Immediately behind the nasal cavity is the upper portion of the muscular pharynx, the _____

3. The portion of the pharynx located behind the mouth is the _____

4. The lowest part of the pharynx, the laryngeal pharynx, opens into the air passageway of the larynx, located toward the front, and into the food pathway, toward the back. This food pathway is called the _____

5. The production of speech is aided by the flow of air from the lungs to vibrate the _____

6. Food is prevented from entering the trachea by closure of the glottis during swallowing. This is accomplished by a leaf-shaped structure called the _____

7. The two main air passageways to the lungs, formed by division of the trachea, are the _____

8. Each bronchus plus the blood vessels and nerves that accompany it enter the lung at a notch or depression called the _____

9. Areas in the major arteries of the chest and neck that regulate breathing according to changes in the composition of the blood are called _____

Group D

    diaphragm   pleura   bronchiole
    glottis   mediastinum   squamous epithelium
    alveoli   hemoglobin   carbon dioxide

1. The space between the two vocal cords is the _____

2. The smallest division of a bronchus is called a(n) _____

3. The air tubes of the lungs finally end in small air sacs called _____

4. The walls of the air sacs in the lungs are very thin to allow for easy passage of gases. These walls are composed of a single layer of _____

5. The heart is located in the space between the lungs called the _____

6. Most of the work of inhalation is done during quiet breathing by the _____

7. The serous membrane around each lung is the _____

8. Most of the oxygen in the blood is carried by _____

9. Bicarbonate ions are formed when a certain gas goes into solution in the blood. This gas is _____

IV. Labeling

For each of the following illustrations, print the name or names of each labeled part on the numbered lines.

1. _____ 6. _____ 11. _____
2. _____ 7. _____ 12. _____
3. _____ 8. _____ 13. _____
4. _____ 9. _____ 14. _____
5. _____ 10. _____ 15. _____
                                                         16. _____
                                                         17. _____
                                                         18. _____

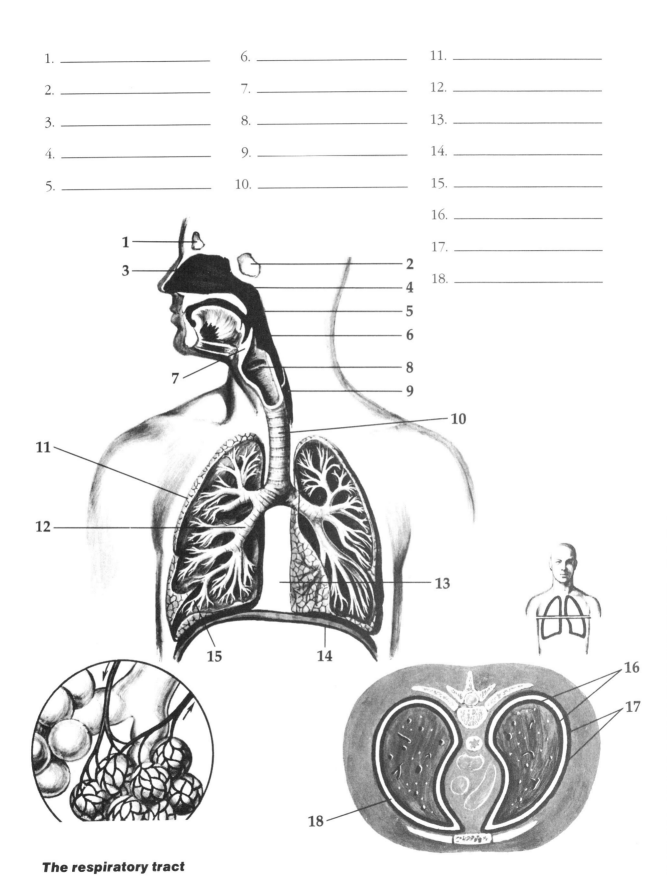

The respiratory tract

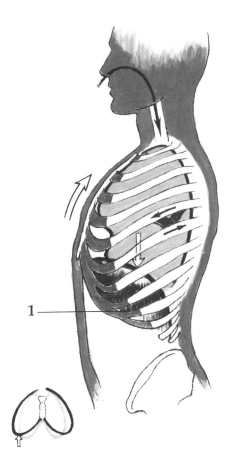 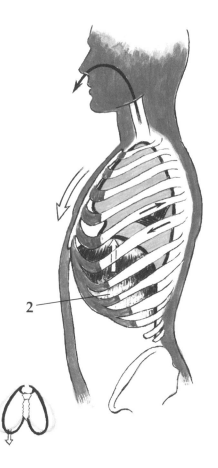

Action of the diaphragm in ventilation

1. _____ 2. _____

V. Completion Exercise

Print the word or phrase that correctly completes each sentence.

1. The nerve that controls the diaphragm is the _____

2. The oxygen that diffuses into the capillary blood of the lungs is bound to a substance in red cell called _____

3. The respiratory control centers are located in the parts of the brainstem called the pons and the _____

4. As bicarbonate ions form in the blood from carbon dioxide, hydrogen ions are also produced. These hydrogen ions tend to make the blood more _____

5. The receptors in the carotid and aortic bodies that are involved in the control of respiration are the _____

6. The region that contains the heart, the large vessels, the esophagus, and the lymph nodes is the _____

VI. Practical Applications

Study each discussion. Then print the appropriate word or phrase in the space provided.

1. Mr. C complained of a severe headache and facial pain. The physician diagnosed the problem as an infection of the air spaces within the cranial bones. These are the _____

2. Young master D, age 5, had a profuse discharge from his nose. There was inflammation of the lining of the nasal cavity. This membrane is made of tissue called _____

3. Ms. F, age 24, complained of difficulty in breathing, partly because of small growths called polyps which were forming between the small lateral projections at the side walls of the nasal cavities. These three projections are the _____

4. An additional problem for Ms. F was a deviated partition between the two nasal spaces. This partition is the _____

5. Ms. A., age 43, complained of hoarseness that interfered with teaching and giving instructions. She had an inflammation of her voice box and windpipe. Another name for the windpipe is _____

6. Mrs. D had been suffering from a severe cold and much coughing. X-rays revealed congestion even in the very smallest subdivisions of the tubes of the bronchial tree. These tiny tubes are called _____

7. Mr. S, age 54, had recovered from an acute attack of pneumonia. X-rays now showed that there was an accumulation of fluid in the right sac that surrounds the lung. This is the _____

16

Digestion

I. Overview

The food we eat is made available to cells throughout the body by the complex processes of **digestion** and **absorption**. These are the functions of the **digestive system**; its components are the **digestive tract** and the **accessory organs**.

The digestive tract, consisting of the **mouth**, the **pharynx**, the **esophagus**, the **stomach**, and the large and small **intestine**, forms a continuous passageway in which ingested food is prepared for utilization by the body and waste products are collected to be expelled from the body. The **liver**, the **gallbladder**, and the **pancreas**, the accessory organs, manufacture various enzymes and other substances needed in digestion. They also serve as storage areas for substances that are released as needed.

Digestion begins in the mouth with the digestion of starch. It continues in the stomach where proteins are digested, and is completed in the small intestine. Most absorption of digested food also occurs in the small intestine through small projections of the lining called **villi**.

The process of digestion is controlled by both nervous and hormonal mechanisms which regulate the activity of the digestive organs and the rate at which food moves through the digestive tract.

II. Topics for Review

A. Components of the digestive tract and the functions of each
B. Accessory organs and their functions
C. The peritoneum
D. The process of digestion
   1. Enzymes involved in digestion
   2. Other substances needed for digestion
E. Absorption
F. Control of the digestive process
   1. Nervous
   2. Hormonal

III. Matching Exercises

Group A

ingestion
tongue
incisors
accessory organs
digestive tract
absorption
digestion
deciduous
molars
premolars

1. The process by which food is converted into substances that may be taken into the cells is known as _____

2. The transfer of digested food to the bloodstream is called _____

3. The structures and organs through which food or its breakdown products pass make up the _____

4. The organs that are needed for digestion but are not part of the digestive tract are the _____

5. The intake of food into the digestive tract is the process of _____

6. Taste sensations can be differentiated by means of special organs of the _____

7. The baby molars are replaced by permanent teeth called bicuspids or _____

8. The grinding teeth located in the back part of the oral cavity are called _____

9. The baby teeth are lost and are therefore described as _____

10. The first baby teeth to appear are the eight cutting teeth or _____

Group B

6-year molars
incisors
canines
third molars
20 teeth
mastication
32 teeth

1. Decay and infection of baby molar teeth may easily spread to the first permanent teeth, the _____

2. The so-called eye teeth are the _____

3. The teeth located in the front part of the oral cavity are the _____

4. It sometimes happens that the jaw is not large enough to accommodate the last teeth to erupt. These teeth are identified as the wisdom teeth or _____

5. By the time the baby is 2 years old he or she should have all the deciduous teeth, which number _____

6. An adult who has a full set of permanent teeth has _____

7. The process of chewing is called _____

Group C

    amylase   esophagus   deglutition
    peristalsis   parotids   pharynx

1. The largest of the salivary glands are the _____

2. The act of swallowing is known as _____

3. The starch-digesting enzyme in saliva is named salivary _____

4. Food is propelled along the digestive tract by the rhythmic motion known as _____

5. In swallowing, the tongue pushes food into the _____

6. The tube that carries food into the stomach is the _____

Group D

    sphincter   omentum   peritoneum
    rugae   villi   mesocolon
    parietal   submucosa   mesentery

1. The serous membrane that lines the abdominal cavity and folds over the abdominal organs is the _____

2. A term that describes the layer of a serous membrane that lines a body cavity is _____

3. Extending downward from the stomach is an apron-like double membrane called the greater _____

4. Nerves, arteries, and other structures supplying the small intestine are found between the two layers of peritoneum called the _____

5. A circular layer of muscle that acts as a valve is a(n) _____

6. The layer of connective tissue beneath the mucous membrane in the wall of the digestive tract is the _____

7. The section of the mesentery around the large intestine is the _____

8. When the stomach is empty, there are many folds in the lining. These folds are called _____

9. The absorbing surface of the small intestine is greatly increased by numerous projections called _____

Group E

pyloric sphincter   soft palate   epiglottis
pharynx   vomiting   enzymes
chyme   cardiac sphincter

1. The uvula hangs from the back of the roof of the oral cavity. This part of the oral cavity roof is the _____

2. During deglutition there is contraction of the muscles of the _____

3. With the muscular contraction that occurs during deglutition, the openings into the air spaces above and below the mouth are closed off by the soft palate and by the _____

4. The structure that guards the entrance into the stomach is called the _____

5. The valve between the distal end of the stomach and the small intestine is the _____

6. In the stomach food is mixed with gastric juice to form _____

7. The active ingredients in gastric juice are hydrochloric acid and _____

8. The feeling of illness known as nausea may be followed by reverse peristalsis, resulting in _____

Group F

jejunum   ileum   sugars
duodenum   proteins   fats
rectum   ileocecal

1. The first part of the small intestine is the _____

2. Lying just beyond the duodenum is the second part of the small intestine, the _____

3. Although bile contains no enzymes, it aids in the digestion of _____

4. The final, and longest, section of the small intestine is the _____

5. Because of its location, the valve between the small and large intestine is described as _____

6. Gastric juice acts mainly to digest _____

7. Among the classes of nutrients that are essential to cell life are carbohydrates, which include _____

8. The sigmoid colon empties into the _____

Group G

glycogen hepat lipase
urea trypsin amylase
albumin

1. A word root that refers to the liver is _____

2. The liver has many essential functions. One of these is the manufacture of a waste product later eliminated by the kidneys. This substance is _____

3. Sugar is stored by the liver and released as simple sugar (glucose) as needed. The form in which sugar is stored is _____

4. One of the many functions of the liver includes the production of plasma proteins such as _____

5. Pancreatic juice contains enzymes that act in various ways on the chyme in the small intestine. Starch is changed to sugar by the pancreatic enzyme _____

6. Fats must be broken down into simpler compounds in order to be absorbed. An enzyme responsible for this breakdown is _____

7. Proteins enter the bloodstream in the form of amino acids. A pancreatic enzyme that splits proteins is called _____

Group H

liver lacteals feces
cecum vermiform appendix small intestine
hydrolysis pepsin colon

1. The important function of absorption is carried out through the numerous villi projecting from the mucosa of the _____

2. Following their absorption into the bloodstream through the capillary walls of the villi, food materials may be stored and released as needed by the _____

3. The enzyme that digests proteins in the stomach is _____

4. Much of fat absorption occurs through the lymphatic capillaries of the villi; they are called _____

5. Food molecules are split in digestion by the addition of water. The chemical name for this process is _____

167

6. Material to be eliminated moves through the ileocecal valve into the beginning of the large intestine. Here it enters a small pouch called the _____

7. The small blind tube attached to the proximal part of the large intestine is called the _____

8. In the large intestine layers of involuntary muscle move the solid waste products toward the rectum. This waste material is called _____

9. The large intestine is commonly called the _____

IV. Labeling

For each of the following illustrations print the name or names of each labeled part on the numbered lines.

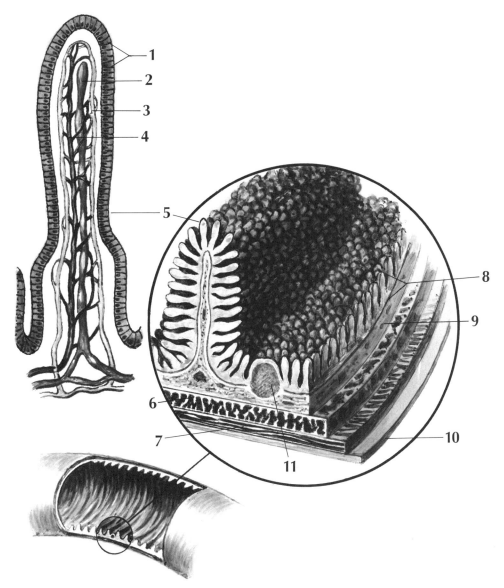

Wall of the small intestine

1. _____
2. _____
3. _____
4. _____
5. _____
6. _____
7. _____
8. _____
9. _____
10. _____
11. _____

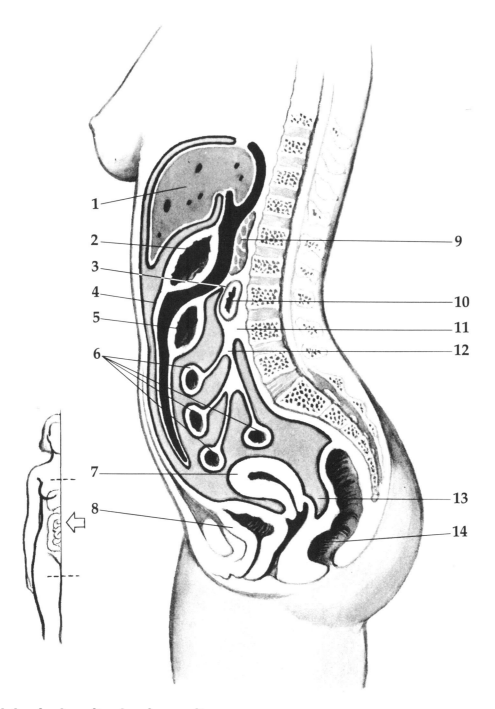

Abdominal cavity showing peritoneum

1. _____   8. _____
2. _____   9. _____
3. _____   10. _____
4. _____   11. _____
5. _____   12. _____
6. _____   13. _____
7. _____   14. _____

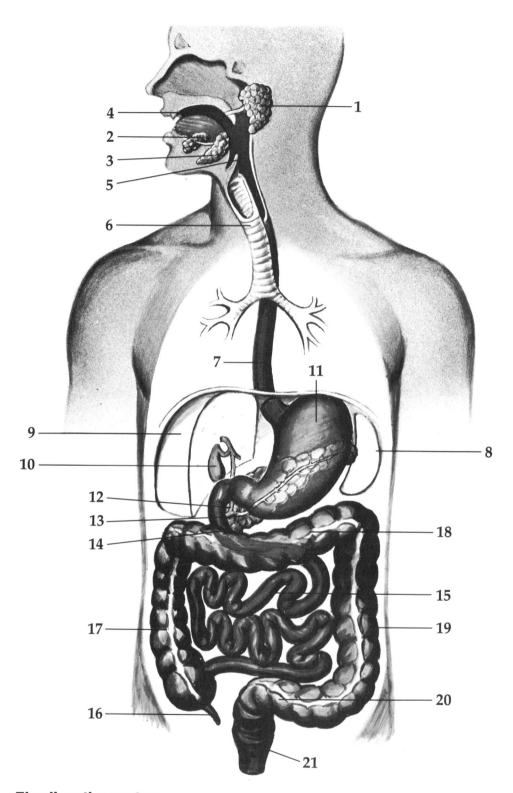

The digestive system

1. _____
2. _____
3. _____
4. _____
5. _____
6. _____
7. _____
8. _____
9. _____
10. _____
11. _____
12. _____
13. _____
14. _____
15. _____
16. _____
17. _____
18. _____
19. _____
20. _____
21. _____

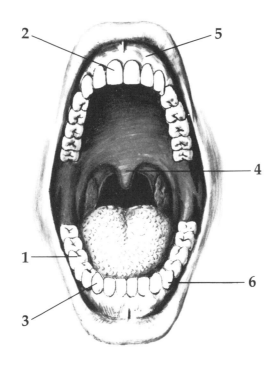

The mouth, showing teeth and tonsils

1. _____
2. _____
3. _____
4. _____
5. _____
6. _____

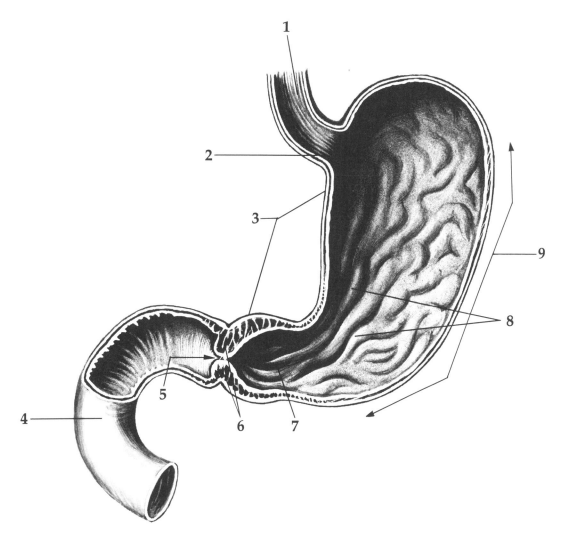

Longitudinal section of the stomach

1. _____
2. _____
3. _____
4. _____
5. _____

6. _____
7. _____
8. _____
9. _____

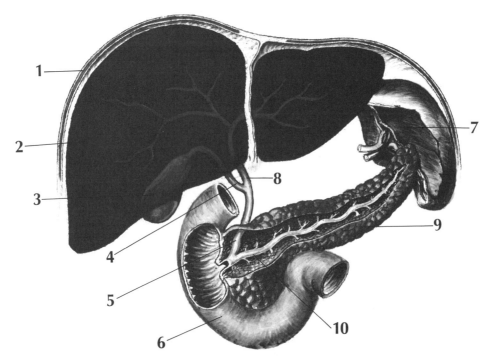

Accessory digestive organs and ducts

1. _____
2. _____
3. _____
4. _____
5. _____
6. _____
7. _____
8. _____
9. _____
10. _____

V. Completion Exercise

Print the word or phrase that correctly completes each sentence.

1. Saliva is produced by three pairs of glands, of which the largest are the ones located near the angles of the jaw, called the _____

2. The salivary glands located under the tongue are the _____

3. One component of gastric juice kills bacteria and thus helps defend the body against disease. This substance is _____

4. Most of the digestive juices contain substances that cause the chemical breakdown of foods without entering into the reaction themselves. These catalytic agents are _____

5. Starches and sugars are classified as _____

6. The lower part of the colon bends into an S-shape, so this part is called the _____

7. A temporary storage section for indigestible and unabsorbable waste products of digestion is a tube called the _____

8. The portion of the distal part of the large intestine called the anal canal leads to the outside through an opening called the _____

9. The muscular sac in which bile is stored to be released as needed is called the _____

VI. Practical Applications

Study each discussion. Then print the appropriate word or phrase in the space provided.

1. Mr. C, age 36, complained of pain in the "pit of the stomach." Ingestion of food seemed to provide some relief. The physician ordered x-ray studies to be done. These studies indicated that the first part of the small intestine was involved. This short section is called the _____

2. Mr. C was a tense man who felt it was important to do well in his business: he worked long hours. Because of this constant stress, his physician suspected that excessive stomach acid was being produced. This acid important in digestion is _____

3. Three-month-old John was brought to the clinic by his mother because he had suffered several bouts of vomiting and could not retain food. The tentative diagnosis was a constriction in the exit valve of the stomach, the _____

4. Mr. K, age 24, complained of pain in his lower abdomen, in the right iliac region. Other symptoms and blood counts indicated that he was suffering from an inflammation of a wormlike appendage of the cecum. The name of this structure is _____

5. Mr. S., age 51, complained of blood in his stool. Investigation revealed a tumor in the last part of the colon. This S-shaped section of the colon is called the _____

6. Mrs. S., age 34, also complained of seeing blood at the time of defecation. Examination revealed enlarged veins called hemorrhoids located in the distal part of the large intestine immediately following the rectum. This leads to the outside through an opening called the _____

7. Mr. G., age 28, had felt under par for some time. Studies showed that Mr. G. had a virus infection of the liver, a disease called hepatitis. Among the many functions of the liver that can be affected by liver disorders is the production of a digestive juice called _____

17

Metabolism, Nutrition, and Body Temperature

I. Overview

The food products that reach the cells following digestion and absorption are used to maintain life. All the physical and chemical reactions that occur within the cells make up **metabolism**, which has two phases: a breakdown phase or **catabolism**, and a building phase or **anabolism**. In catabolism food is oxidized to yield energy for the cells in the form of ATP. This process, termed **cellular respiration**, occurs in two steps: the first is anaerobic (does not require oxygen) and produces a small amount of energy; the second is aerobic (requires oxygen). This second step occurs within the mitochondria of the cells. It yields a large amount of the energy contained in the food plus carbon dioxide and water.

By the various pathways of metabolism, the breakdown products of food can be built into substances needed by the body. The **essential** amino acids and fatty acids cannot be manufactured internally and must be taken in with the diet. Minerals and vitamins are also needed in the diet for health. Since ingested food is the source of all nourishment for the body, a balanced diet should be followed and "food fads" should be avoided.

The rate at which energy is released from food is termed the **metabolic rate**. It is affected by many factors including age, size, sex, activity, and hormones. Some of the energy in food is released in the form of heat, which serves to maintain body temperature. A steady temperature of about 37°C (98.6°F) is maintained by several mechanisms.

Heat production is greatly increased during periods of increased muscular or glandular activity. Most heat loss occurs through the skin, with a smaller loss by way of the respiratory system and the urine and feces. The **hypothalamus** of the brain maintains the normal temperature in response to the temperature of the blood and information received from temperature receptors in the skin. Regulation occurs through vasodilation and vasoconstriction of the surface blood vessels, activity of the sweat glands, and muscle activity.

II. Topics for Review
A. Metabolism
   1. Catabolism
   2. Anabolism
B. Nutrition
   1. Minerals and vitamins
   2. Essential amino acids and fatty acids
   3. Balanced diet
C. Metabolic rate
D. Body temperature
   1. Heat production
   2. Heat loss
   3. Temperature regulation
      a. Role of the hypothalamus
      b. Normal body temperature

III. Matching Exercises
Matching only within each group, print the answers in the spaces provided.

Group A
glycogen   BMR   ATP
anaerobic   kilocalorie   catabolism
glycerol   anabolism   mitochondria

1. The breakdown of glucose for energy is an example of _____

2. The first phase of cellular respiration does not require oxygen and so is described as _____

3. A compound that stores energy in the cell _____

4. The aerobic steps of metabolism occur within the cell organelles called _____

5. The unit used to measure the energy in foods is the _____

6. The storage form of glucose is _____

7. The manufacture of proteins from amino acids within the cell is an example of _____

8. A component of fats is _____

9. The amount of energy needed to maintain life functions while the body is at rest is the _____

Group B
aerobic   minerals   essential
oxidation   glucose   enzymes
saturated   legumes

1. The main energy food for the cells is _____

2. An amino acid that must be taken in as part of the diet is described as _____

3. The steps of metabolism that release most of the energy from food require oxygen and are termed _____

4. The breakdown of food for energy involves the chemical process of _____

5. The catalysts of metabolic reactions are the _____

6. In addition to fats, proteins, and carbohydrates, the body requires vitamins and _____

7. Fats that are solid at room temperature and are mostly from animal sources are termed _____

8. Peas and beans are classified as _____

Group C

calciferol	niacin	calcium
B_1	A	B_{12}
C	iron	potassium

1. Beriberi can be prevented by the intake of adequate amounts of thiamine, also known as vitamin _____

2. A constituent of the oxygen-carrying compound hemoglobin is the element _____

3. The vitamin that prevents dry, scaly skin and night blindness is _____

4. The vitamin required for normal bone formation is called vitamin D or _____

5. An element required for normal nerve and muscle activity is found in certain foods in the form of a mineral salt. It is _____

6. Another name for ascorbic acid is vitamin _____

7. A strict vegetarian who eats no eggs or dairy products must be careful to avoid anemia due to a lack of vitamin _____

8. Pellagra may result from a lack of _____

9. A mineral needed for proper bone development and that is found in dairy products and vegetables is _____

Group D

oxygen
homeostasis
glands

evaporation
constrict
skin

blood
hypothalamus
insulation

1. Body heat is produced by combination of food products with _____

2. The largest amount of heat is produced in the body by muscles and _____

3. The tendency of body processes to maintain a constant state is called _____

4. Heat is distributed throughout the body by way of the _____

5. The body possesses several means of ridding itself of heat; the largest part of this loss occurs through the _____

6. Clothing and subcutaneous fat represent types of _____

7. The chief heat-regulating center, located in the brain, is the _____

8. If too much heat is being lost from the body, the blood vessels in the skin are caused to _____

9. The amount of humidity in the air has an effect on the rate of heat loss by _____

IV. Completion Exercise

Print the word or phrase that correctly completes each sentence.

1. All the physical and chemical reactions that sustain life together make up _____

2. The process of oxidizing food within the cell for energy is termed _____

3. Organic substances needed in small amounts in the diet are the _____

4. A gland important in the control of the metabolic rate is the _____

5. While most heat loss occurs through the skin, an appreciable amount is also lost in the urine and feces and by way of the _____

6. The most important heat-regulating center is a section of the brain called the _____

7. Shivering is a way of increasing body heat by increasing the activity of the _____

8. The normal range of body temperature in degrees Celsius is _____

9. The formula for converting Fahrenheit temperatures to Celsius is _____

10. Practice changing Fahrenheit to Celsius. Show the figures for changing 50° and 70°F to Celsius. _____

11. Practice changing Celsius to Fahrenheit. Show the figures for changing 10° and 25°C to Fahrenheit. _____

V. Practical Applications

Study each discussion. Then print the appropriate word or phrase in the space provided.

Group A

1. Mrs. S, age 76, was hospitalized for a fracture of the femur caused by a fall. Her physician suspected that the break had resulted from a general weakening of the bones due to osteoporosis. This disorder, common in elderly women, is caused by a number of factors, including the dietary lack of a mineral found in dairy products. This mineral is _____

2. Mr. C, age 78, was accompanied by his daughter to visit his family physician. His daughter was concerned about his general state of health and marked weight loss within several months after the death of his wife. The doctor asked that Mr. C keep a record of his food intake for 2 weeks. Review of this record suggested that he was not eating properly and was suffering from a general lack of proper nutrients in his diet. The doctor described his condition as one of borderline _____

3. Young Mr. N, age 17, had placed himself on a strict vegetarian diet that included no animal products. He was not careful in planning his meals, however, and his family soon began to notice his loss of appetite, irritability, and susceptibility to disease. The school dietitian, when questioned by his mother, suggested that he was not getting the right balance of proteins, especially the building blocks of proteins that must be taken in with the diet, the _____

4. When Ms. R, age 15, went for her regular dental examination, the dentist noticed that her gums bled easily and that she had small cracks at the corners of her mouth. Brief questioning suggested that because of a lack of fruits and vegetables in her diet she was suffering from a lack of vitamins, especially vitamin C and a group of vitamins that includes thiamine and riboflavin, the _____

Group B

A physician working in a desert area of southeastern California saw a variety of cases during the course of a day.

1. A 6-year-old patient appeared apathetic and tired. His face was flushed and hot. On taking his temperature the nurse found it to be 105°F. The physician took the child's history and examined him, then instructed his mother to give the child cool sponge baths and administer the prescribed medication. The cool water sponging would aid in reducing the temperature by the process of _____

2. Men working on a construction project complained of tiredness and nausea. They felt better after resting in the shade and drinking water and fruit juices. Their fluid losses had been caused mainly by increased activity of the _____

3. Mr. K, age 69, had been working in his garden. The day was sunny and hot, but Mr. K neglected to protect his bald head by wearing a hat. He began to feel dizzy and faint. His wife noted that his face was very flushed because the blood vessels in his skin had been caused to _____

18
The Urinary System and Body Fluids

I. Overview

The urinary system comprises two **kidneys**, two **ureters**, one **urinary bladder**, and one **urethra**. This system is thought of as the body's main excretory mechanism; it is, in fact, often called the **excretory system**. The kidney, however, performs other essential functions; it aids in maintaining water and electrolyte balance and in regulating the acid–base balance (pH) of body fluids. The kidneys also secrete a hormone that stimulates red blood cell production and another hormone that acts to increase blood pressure.

The functional unit of the kidney is the **nephron**. It is the nephron that produces **urine** from substances filtered out of the blood through a cluster of capillaries, the **glomerulus**. Oxygenated blood is brought to the kidney by the **renal artery**. The arterial system subdivides as it branches through the kidney until the smallest vessel, the **afferent arteriole**, carries blood into the glomerulus. Blood leaves the glomerulus by means of the **efferent arteriole** and eventually leaves the kidney by means of the **renal vein**. Before blood enters the venous network of the kidney, exchanges occur between the filtrate and the blood through the **peritubular capillaries** that surround each nephron.

The composition of intracellular and extracellular fluids is an important factor in homeostasis. These fluids must have the proper levels of electrolytes and must be kept at a constant pH.

Other mechanisms, in addition to kidney function, that help to regulate the composition of body fluids are the thirst mechanism, hormones, buffers, and respiration. Normally, the amount of fluid taken in with food and beverages equals the amount of fluid lost through the skin, and the respiratory, digestive, and urinary tracts. The normal pH of body fluids is a slightly alkaline 7.4.

II. Topics for Review

A. The urinary system
   1. Kidneys
   2. Ureters
   3. Bladder
   4. Urethra
B. Renal function
   1. Glomerular filtration
   2. Tubular reabsorption
   3. Tubular secretion
   4. Concentration of the urine
      a. Role of ADH
C. Urine
D. Regulation of body fluids
   1. Electrolytes
      a. Role of hormones
   2. Concentration
      a. Loss and gain of water
      b. The thirst mechanism
   3. Acid–base balance (pH)
      a. Buffers
      b. Kidney function
      c. Respiration

III. Matching Exercises

Matching only within each group, print the answers in the spaces provided.

Group A

digestive system   nephrons   elimination
respiratory system   adipose capsule   excretion
retroperitoneal space   urine   fibrous capsule

1. Removal of waste products from the body is called _____

2. By contrast, the actual emptying of the hollow organs in which waste substances have been stored is referred to as _____

3. Other systems besides the urinary system perform excretory functions. To mention one example, bile is excreted by the _____

4. The system regulating excretion of carbon dioxide and water is the _____

5. The urinary system excretes water, nitrogenous waste products, and salts, all of which are contained in the _____

6. The functional units of the kidney are the microscopic _____

7. The membranous connective tissue structure that is normally loosely adherent to the kidney itself is called the _____

8. The area behind the peritoneum that contains the pancreas, the duodenum, and the two kidneys is referred to as the _____

9. The circle of fat that helps to support the kidney is called the _____

Group B

collecting tubules
renal pelvis
cortex
glomerulus
epithelium
Bowman's capsule
convoluted tubules
filtration
urea
reabsorption

1. The cluster of capillaries located at the proximal end of the nephron is the _____

2. Materials that have passed through the capillary walls enter the first part of the nephron, the _____

3. The longest sections of the nephrons are the _____

4. Useful substances that have entered the nephron are sent back to the bloodstream by a process of _____

5. The glandular kidneys are made up mainly of _____

6. The process by which substances leave the glomerulus and enter Bowman's capsule is _____

7. The outer region of the kidney is the _____

8. Within the medulla the distal convoluted tubules of the nephrons come together and empty into the _____

9. The upper end of the ureter is a funnel-shaped basin that receives urine. It is called the _____

10. As body cells use protein, nitrogenous waste products are produced; the chief such product is _____

Group C

electrolytes
glucose
calyces
internal sphincter
urethra
organic
peristalsis
hilus
hydrogen ions

1. Urine is moved along the ureter from the kidneys to the bladder by the rhythmic contraction known as _____

2. Near the bladder outlet are circular muscle fibers that contract to prevent emptying. They form what is known as the _____

3. The tube that carries urine from the bladder to the outside is the _____

4. Because nitrogen waste products originate from living organisms, they are said to be _____

5. Mineral salts contained in urine are classified as _____

6. Diabetes mellitus may be suspected if a test of the urine shows the presence of the simple sugar _____

7. The area where the artery, the vein, and the ureter connect with the kidney is known as the _____

8. Tube-like extensions that project from the renal pelvis into the kidney tissue serve to increase the area for collection of urine. These extensions are called _____

9. The kidney helps prevent conditions of excessive alkalinity or acidity by regulating the body fluid concentration of _____

Group D

erythropoietin   dilute   concentrated
juxtaglomerular cells   renin   nitrogen
reabsorption   secretion   specific gravity

1. The amount of dissolved substances in the urine is indicated by its _____

2. The kidney releases a hormone-like substance that acts to produce a stimulator of the red bone marrow. This stimulator is called _____

3. Renin and other hormones are produced in the kidney by the _____

4. Certain cells in the kidney produce a hormone that activates a blood protein, which in turn induces an increase in blood pressure. This hormone is called _____

5. A specific gravity of 1.002 would indicate that the urine is very _____

6. Urea, uric acid, and creatinine are waste products that are derived from proteins; all contain the element _____

7. The term for the process by which useful substances are returned to the tissue fluid and the blood is _____

8. If the specific gravity of urine is approximately 1.040, it contains considerable amounts of solutes, and is said to be _____

9. Urine is more acid than blood because the renal tubule actively moves hydrogen ions from the blood into the tubule to be excreted. This active process is called tubular _____

Group E

negative
extracellular
intracellular
interstitial
cation
parathyroid hormone
buffers
aldosterone
acidity

1. Sodium carries a positive electric charge, and therefore is a(n) _____

2. Phosphate carries electric charges that are _____

3. Fluid within the body cells is designated _____

4. Plasma is in the compartment classified as _____

5. The water located in the microscopic spaces between cells is designated _____

6. A hormone produced by the adrenal cortex that promotes the reabsorption of sodium is _____

7. A hormone that causes the kidneys to reabsorb calcium is _____

8. Exhalation of carbon dioxide is one means of reducing the blood's _____

9. Body fluids are maintained at a constant pH partly by the action of _____

IV. Labeling

For each of the following illustrations print the name or names of each labeled part on the numbered lines.

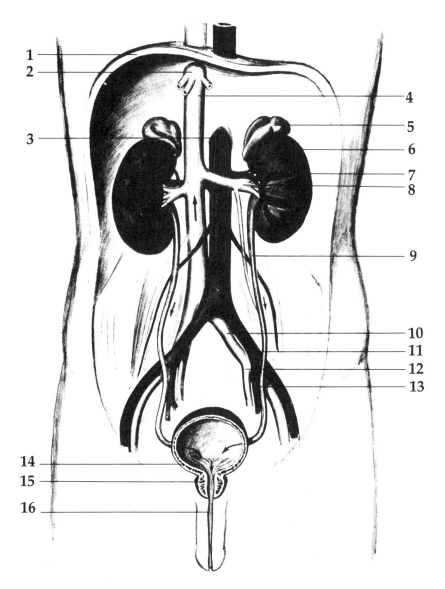

Urinary system with blood vessels

1. _____
2. _____
3. _____
4. _____
5. _____
6. _____
7. _____
8. _____
9. _____
10. _____
11. _____
12. _____
13. _____
14. _____
15. _____
16. _____

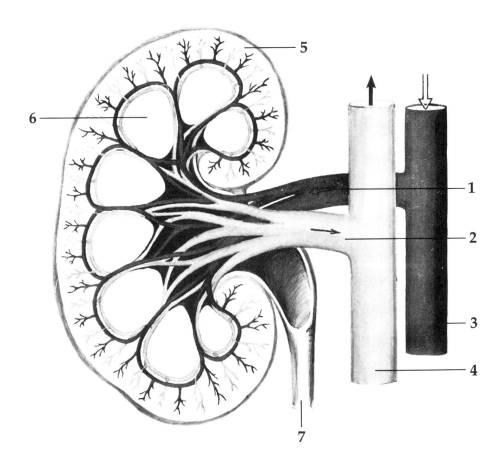

Blood supply and circulation of kidney

1. _____
2. _____
3. _____
4. _____
5. _____
6. _____
7. _____

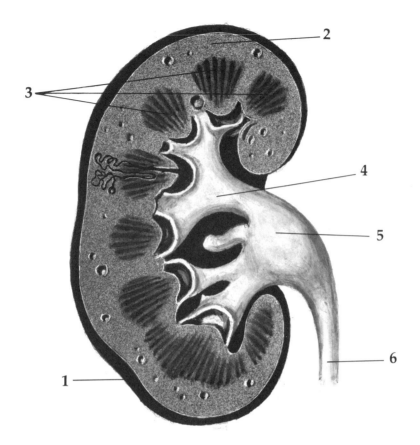

Kidney's internal structure

1. _____   4. _____

2. _____   5. _____

3. _____   6. _____

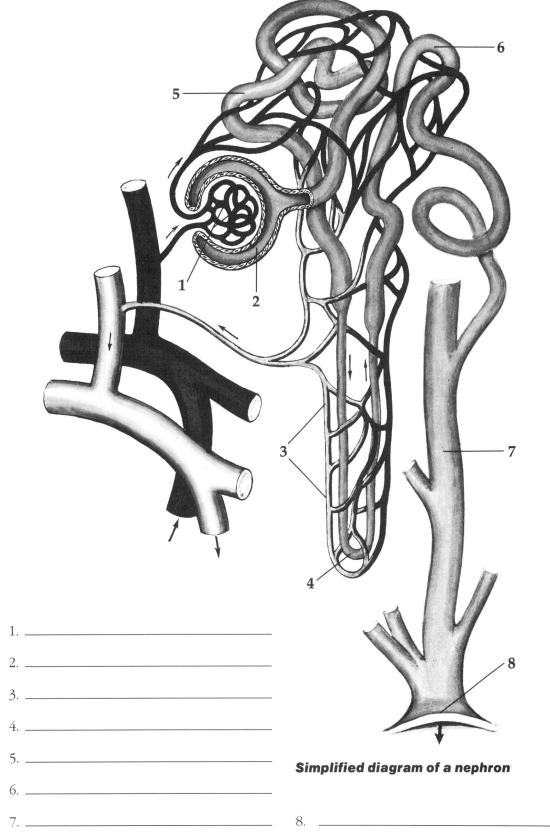

1. _____
2. _____
3. _____
4. _____
5. _____
6. _____
7. _____   8. _____

Simplified diagram of a nephron

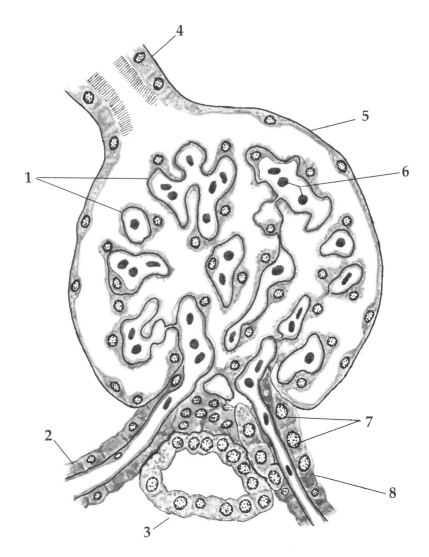

Structure of the juxtaglomerular apparatus

1. _____ 5. _____

2. _____ 6. _____

3. _____ 7. _____

4. _____ 8. _____

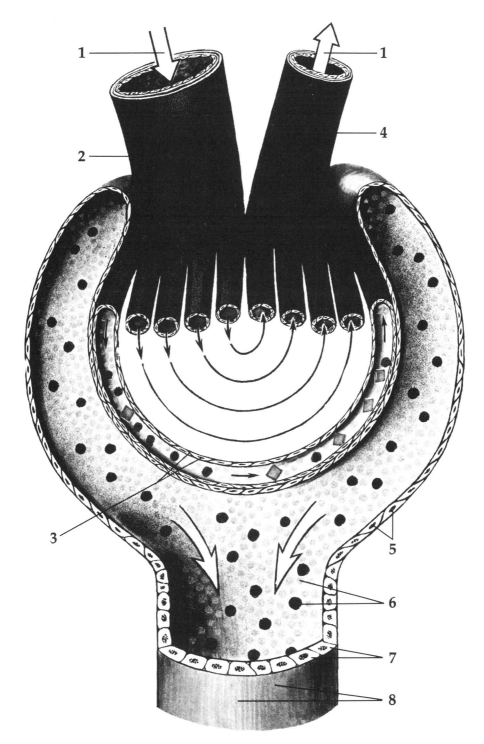

Diagram showing filtration process during formation of urine

1. _____ 5. _____

2. _____ 6. _____

3. _____ 7. _____

4. _____ 8. _____

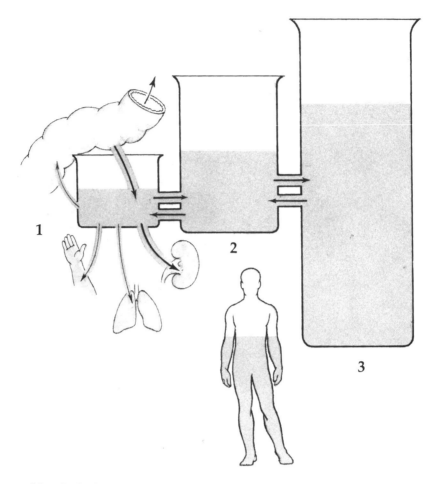

Distribution of body fluids

1. _____   3. _____

2. _____

V. Completion Exercise

Print the word or phrase that correctly completes each sentence.

1. Because of the oblique direction of the last part of each ureter through the lower bladder wall, compression of the ureters by the full bladder prevents _____

2. When the bladder is empty, its lining is thrown into the folds known as _____

3. The vessel that carries oxygenated blood to the kidney is the _____

4. In the male, two ducts that carry sex cells join the first part of the urethra as it passes through the _____

5. Urine consists mainly of _____

6. Water balance is partly regulated by a thirst center located in a region of the brain called the _____

7. The hormone from the pituitary that regulates water reabsorption in the kidney is _____

VI. Practical Applications

Study each discussion. Then print the appropriate word or phrase in the space provided.

1. Mrs. K was suffering from cystitis, a bladder infection. Studies indicated that there was relaxation of the pelvic floor, causing stagnation of urine in the bladder, and corrective surgery was planned. In preparation for this, a tube (catheter) was inserted into the external opening, the _____

2. Mr. K, age 61, required several studies to determine the cause of obstruction of his urinary tract. One of the studies revealed that there was an obstruction at the bladder neck, a disorder that is fairly common in men of his age. The obstruction was caused by enlargement of the gland through which the first part of the urethra passes. This is the _____

3. Mr. R was suffering from nephritis, or inflammation of the kidney. In this disorder there is destruction of the functional units of the kidney, called _____

4. A test of Mr. R's blood revealed an abnormally high content of the chief nitrogen waste product called _____

5. Mr. J, age 58, complained of difficulty in emptying his bladder. examination (with a cystoscope) showed an enlargement of the prostate gland. The male sex cells are carried into the first part of the excretory tube called the _____

6. Investigation of Mr. J's urinary problem included the passage of catheters up through the urethra and urinary bladder, and finally through the ureters into the kidney basin, also called the _____

7. Mr. J had difficulty emptying his bladder. The process of emptying the bladder is called urination or _____

8. The urine was examined in the case of each of these patients. The main constituent of urine, about 95%, is _____

19

Reproduction

I. Overview

Reproduction is the process by which life continues. Human reproduction is **sexual**, that is, it requires the union of two different **germ cells** or **gametes**. (Some simple forms of life can reproduce without a partner in the process of **asexual** reproduction.) These germ cells, the **spermatozoon** in males and the **ovum** in females, are formed by **meiosis**, a type of cell division in which the chromosome number is reduced to one half. When fertilization occurs and the gametes combine, the original chromosome number is restored.

The reproductive glands or **gonads** manufacture the gametes and also produce hormones. These activities are continuous in the male but cyclic in the female. The male gonad is the **testis**. The remainder of the male reproductive tract consists of passageways for transport of spermatozoa; the male organ of copulation, the **penis**; and several glands that contribute to the production of **semen**. The female gonad is the **ovary**. The ovum released each month at the time of **ovulation** travels through the **oviducts** to the **uterus**, where the egg, if fertilized, develops. If no fertilization occurs, the ovum, along with the built-up lining of the uterus, is eliminated through the **vagina** as the **menstrual flow**.

Reproduction is under the control of hormones from the **anterior pituitary** which, in turn, is controlled by the **hypothalamus** of the brain. These organs respond to **feedback** mechanisms which maintain proper hormone levels.

Pregnancy is the period of about 9 months during which a fertilized ovum develops, first as an **embryo** and then as a **fetus**. During this period the developing offspring is nourished and maintained by the **placenta**, formed from tissues of both the mother and the embryo.

Aging causes changes in both the male and female reproductive systems. A gradual decrease in male hormone production begins as early as age 20 and continues throughout life. In the female a more sudden decrease in activity occurs between ages 45 to 55 and ends in the **menopause**, the cessation of menstruation and of the child-bearing years.

II. Topics for Review

A. Formation of the germ cells
   1. The spermatozoon
   2. The ovum
   3. Meiosis
B. The male reproductive tract
   1. Testes
   2. Ducts
   3. Penis
   4. Glands
      a. Semen
C. The female reproductive tract
   1. Ovaries
   2. Oviducts
   3. Uterus
   4. Vagina
   5. Vulva
D. Hormonal control of reproduction
   1. Pituitary
   2. Hypothalamus
   3. Feedback
E. The menstrual cycle
   1. Menopause
F. Pregnancy
   1. Fertilization
   2. Embryo
   3. Fetus
   4. Placenta
   5. Childbirth
   6. Lactation

III. Matching Exercises

Matching only within each group, print the answers in the spaces provided.

Group A

ovum	spermatozoa	epididymis
gonads	ovary	sexual
testis	asexual	

1. The specialized sex cells in the male are called   _____

2. Since the simplest forms of life require no partner in order to reproduce, they are said to be   _____

3. The sex glands are also called the   _____

4. The specialized sex cell of the female is the   _____

5. The female gonad is also known as the   _____

6. The male gonad is the   _____

7. The spermatozoa mature and become motile within a temporary storage area, a 20-foot tube, the _____

8. Most animal species are differentiated into males and females; within these groups reproduction is said to be _____

Group B

ejaculatory duct   ductus deferens   FSH
spermatic cord   gametes   urethra
acrosome   penis   semen

1. A term used in referring to the male and female sex cells is _____

2. The straight upward extension of the epididymis is the _____

3. The combination of ductus deferens, nerves, and blood and lymph vessels that extends from the scrotum and testes on each side is named the _____

4. The ductus deferens on each side is joined by the duct from the seminal vesicle to form a tube that carries spermatozoa through the prostate. This tube is the _____

5. In males, a single tube conveys urine and semen to the outside. This tube is the _____

6. The external genitalia of the male include the scrotum and the _____

7. In ejaculation, a mixture of spermatozoa and secretions is expelled. It is called _____

8. A caplike covering over the head of the spermatozoon that aids in penetration of the ovum is the _____

9. Sertoli cells are stimulated by _____

Group C

penis   Cowper's glands   scrotum
testosterone   seminiferous   seminal vesicles
ejaculation

1. In the male, the longest part of the urethra extends through the spongelike _____

2. The bulbourethral glands, which are pea-sized organs found in the pelvic floor tissues below the prostate gland of the male, are also known as _____

3. The testes are normally located in a sac that is suspended between the thighs. This sac is the _____

4. The bulk of the tissue of the testes is arranged in tubules. These tubules are described by the term _____

5. Groups of cells located between the tubules of the testes are responsible for the secretion of the male hormone named _____

6. Behind the urinary bladder in the male are two tortuous muscular tubes with glandular linings. These are the _____

7. Semen is expelled by a series of muscular contractions called _____

Group D

vulva	vagina	fallopian tubes
ovaries	ovarian follicles	uterus
ovulation	Bartholin's glands	fimbriae

1. Two structures, made of peritoneum and called the broad ligaments, serve as anchors for the uterus and _____

2. The sacs within which the ova mature are called the _____

3. The rupture of an ovarian follicle permits an ovum to be discharged from the ovary surface. This is called _____

4. The mature ovum travels from the region of the ovary into the oviducts or _____

5. A current in the peritoneal fluid sweeps the ovum into the oviduct. This current is produced by the fringelike _____

6. Before birth the fetus grows in a muscular organ located between the urinary bladder and the rectum. This organ is the _____

7. The greater vestibular glands, situated above and to each side of the vaginal opening are also known as _____

8. Connecting the uterus with the outside is the lower part of the birth canal, the _____

9. The labia, the clitoris, and related structures comprise the external parts of the female reproductive system which are called the _____

Group E

menopause	cervix	corpus luteum
external genitalia	fundus	endometrium
corpus	fornices	estrogen

1. Located above the level of the tubal entrances is the small rounded part of the uterus called the _____

2. The upper part of the uterus is the largest part; it is called the body, or _____

3. The necklike part of the uterus dips into the upper vagina; this necklike part is called the _____

4. The specialized tissue that lines the uterus is known as _____

5. The cervix dips into the upper vagina so that a circular recess is formed; this gives rise to the spaces known as _____

6. The vulva is also called the _____

7. Cessation of ovarian activity brings about the period of life known as _____

8. Two ovarian hormones are involved in preparing the endometrium for pregnancy, and both are carried by the bloodstream to the uterus. Preparation is initiated by hormones produced by the follicle as the ovum matures. These hormones as a group are called _____

9. After ovulation, the ruptured follicle becomes the _____

Group F

parturition   umbilicus   embryo
fetus   placenta   cervix
afterbirth   vernix caseosa   progesterone
amniotic sac

1. The endometrium is prepared for the fertilized ovum by a hormone produced by the corpus luteum. This hormone is _____

2. Serving as the organ for nutrition, respiration, and excretion for the embryo is a flat, circular structure called the _____

3. Following fertilization of an ovum and until the end of the third month, the developing organism is called the _____

4. From the end of the third month until birth the developing organism is known as the _____

5. The fetus is protected by a fluid contained in the _____

6. Nature provides various protective mechanisms for the fetus. The cheesy material that protects the skin is known as _____

7. The process of giving birth to a child is described by the term labor or _____

8. A small part of the umbilical cord remains attached to the navel for a few days following birth. The scientific name for the navel is the _____

9. Normally, within half an hour after the child is born, the placenta together with the membranes of the amniotic sac and most of the umbilical cord are expelled as the _____

10. In the first stage of childbirth, the regular contractions of the uterus bring about widening of the opening of the _____

Group G

menstrual flow
placental lactogen
luteinizing hormone
follicle stimulating hormone
posterior fornix
colostrum
chorionic gonadotropin
umbilical cord

1. The sac containing the ovum develops under the influence of a hormone from the pituitary called _____

2. Midway during the menstrual cycle the ovarian follicle ruptures under the effects of a hormone from the pituitary called _____

3. Oxygen and nutrients are brought to the fetus and waste products are removed from the fetus through blood vessels contained in the _____

4. Without hormones to support growth, the thickened endometrium is shed, producing the _____

5. A hormone produced by embryonic cells that maintains the corpus luteum early in pregnancy is _____

6. Several hormones are secreted during pregnancy by the placenta, including estrogen, progesterone, and _____

7. The peritoneal cavity of the female is deepest behind the upper vaginal canal. This means that there is a thin wall separating the lower abdominal cavity from the upper vaginal canal. This dorsal space in the upper vagina is called the _____

8. The first mammary gland secretion to appear is called _____

IV. Labeling

For each of the following illustrations, print the name or names of each labeled part on the numbered lines.

1. _____
2. _____
3. _____
4. _____
5. _____
6. _____
7. _____
8. _____
9. _____
10. _____
11. _____
12. _____
13. _____
14. _____

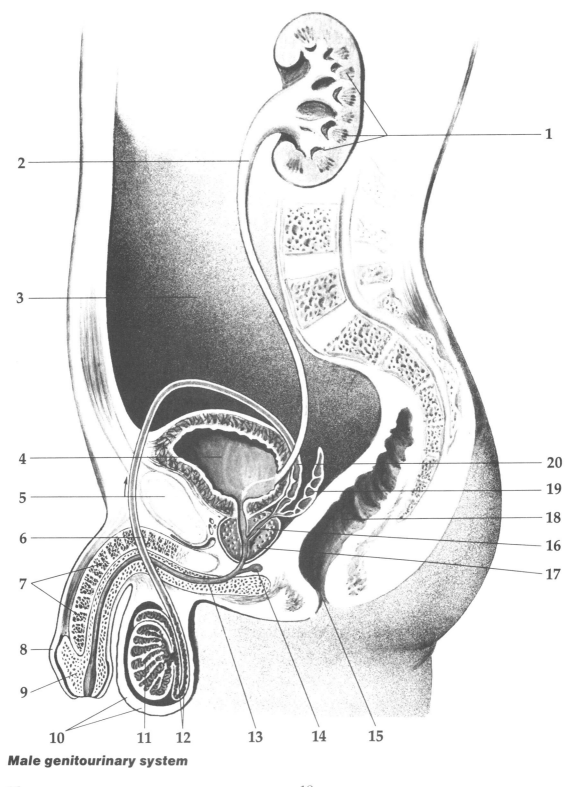

Male genitourinary system

15. _____   18. _____

16. _____   19. _____

17. _____   20. _____

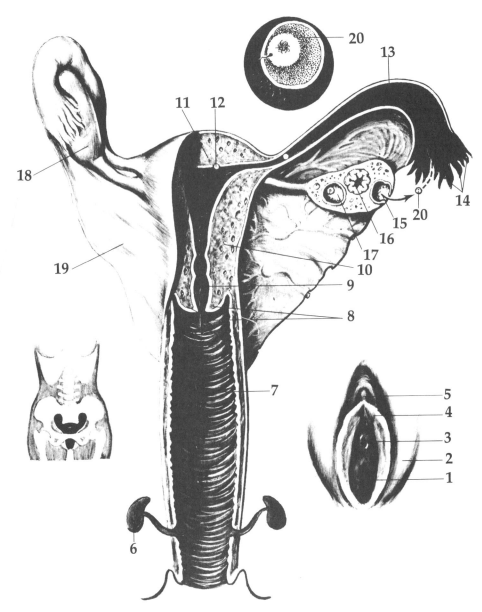

Female reproductive system

1. _____
2. _____
3. _____
4. _____
5. _____
6. _____
7. _____
8. _____
9. _____
10. _____
11. _____
12. _____
13. _____
14. _____
15. _____
16. _____
17. _____
18. _____
19. _____
20. _____

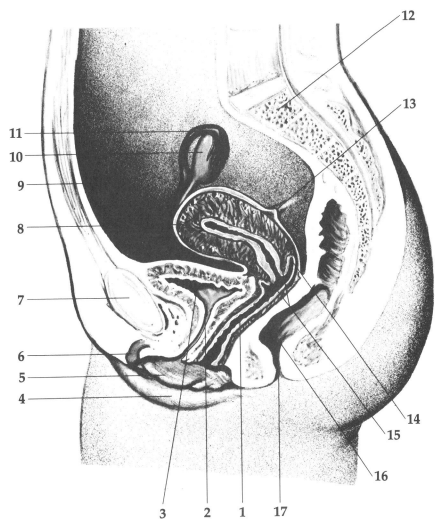

Female reproductive system, sagittal section

1. _____
2. _____
3. _____
4. _____
5. _____
6. _____
7. _____
8. _____
9. _____
10. _____
11. _____
12. _____
13. _____
14. _____
15. _____
16. _____
17. _____

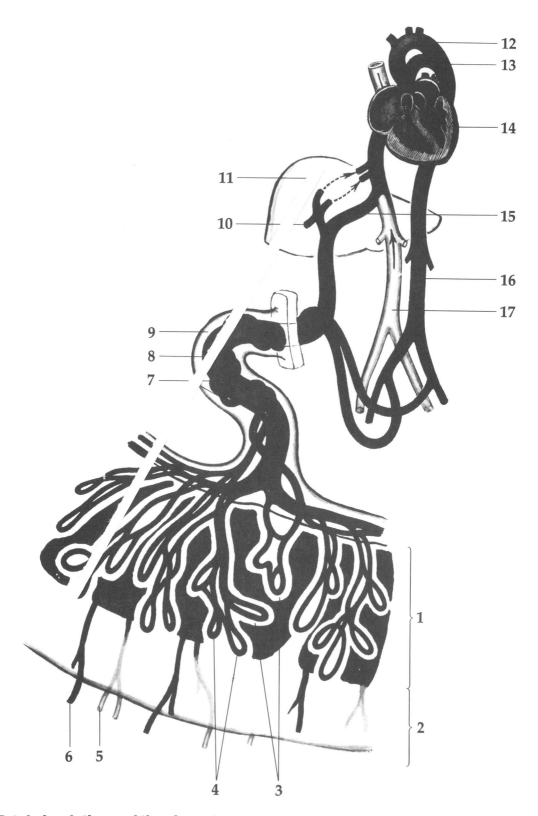

Fetal circulation and the placenta

1. _____   10. _____
2. _____   11. _____
3. _____   12. _____
4. _____   13. _____
5. _____   14. _____
6. _____   15. _____
7. _____   16. _____
8. _____   17. _____
9. _____

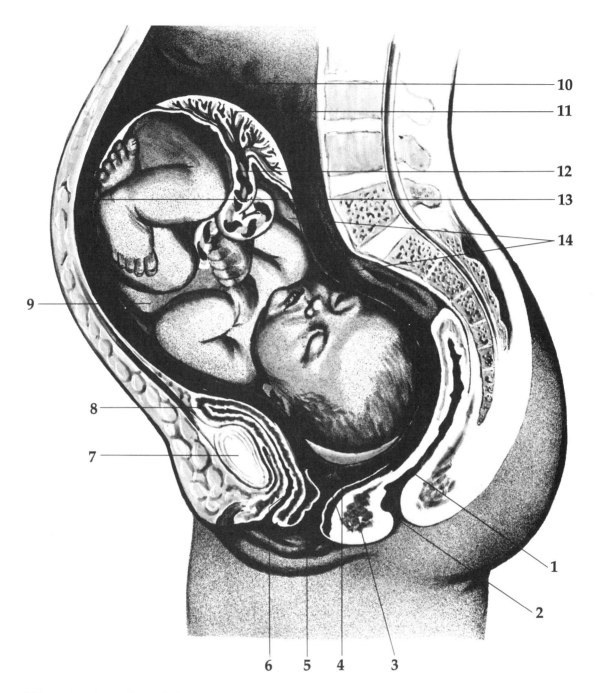

Midsagittal section of the pregnant uterus

1. _____
2. _____
3. _____
4. _____
5. _____
6. _____
7. _____
8. _____

9. _____   12. _____
10. _____   13. _____
11. _____   14. _____

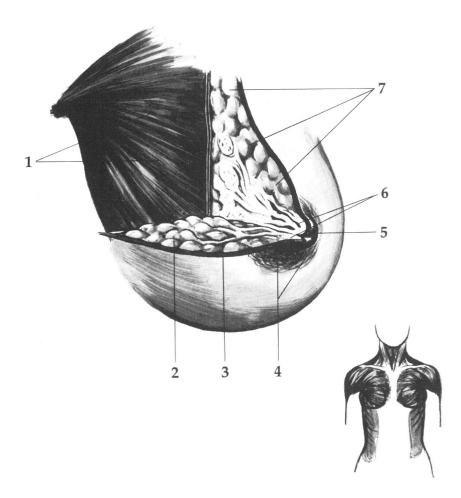

Section of breast

1. _____   5. _____
2. _____   6. _____
3. _____   7. _____
4. _____

V. Completion Exercise

Print the word or phrase that correctly completes each sentence.

1. The process of cell division that reduces the chromosome number by half is _____

2. The individual spermatozoon is very motile. It is able to move toward the ovum by the action of its _____

3. By the end of the first month of embryonic life, the beginnings of the extremities may be seen. These are four small swellings called _____

4. The cell formed by the union of a male sex cell and a female sex cell is called a _____

5. The science that deals with the development of the embryo is called _____

6. The bag of waters is a popular name for the membranous sac that encloses the fetus. The clear liquid that is released during labor is called _____

7. About once in every 80 to 90 births twins are born. Some of these twins occur as a result of two different ova being fertilized by two spermatozoa. Such twins are said to be _____

8. Some twins develop from a single zygote formed from a single ovum that has been fertilized by a single spermatozoon. The embryonic cells separate during early stages of development. These twins are described as _____

9. The mammary glands of the female provide nourishment for the newborn through the secretion of milk; this is a process called _____

10. The hormone that causes ovulation in females is the same hormone that stimulates cells in the testes to produce testosterone. In females this hormone is called luteinizing hormone (LH); in males it is called _____

11. The passage of the fetus through the cervical canal and the vagina to the outside takes place during the period of labor termed the _____

12. The use of artificial methods to prevent fertilization or implantation of the fertilized ovum is called _____

VI. Practical Applications

Study each discussion. Then print the appropriate word or phrase in the space provided.

20

Heredity

I. Overview

The scientific study of heredity is less than 2 centuries old. During the last few decades there have been brilliant and illuminating findings particularly related to the chromosomes and DNA; nevertheless, many mysteries remain and there is much more to be studied. Gregor Mendel was the first person to put the study of heredity on a scientific footing. He called attention to the independent units of hereditary influence, which he called *factors* and which we now call **genes**.

There are many thousands of genes in each cell nucleus. Each chromosome in the nucleus is composed of a complex molecule, DNA, and the genes are parts of this molecule. The chromosomes containing the genes are passed on to offspring through the **germ cells** formed in each parent by the process of **meiosis**. Genes direct the formation of **enzymes**, which in turn make possible the chemical reactions of metabolism. Some human traits are determined by a single pair of genes (one gene from each parent), but most are controlled by multiple pairs of genes acting together.

Genes may be classified as **dominant** or **recessive**. If one parent contributes a dominant gene, then all those offspring who receive that gene will show the trait. Traits carried by recessive genes may remain hidden for generations and be revealed only if they are contributed by both parents. A **mutation** is a change in a gene or a chromosome which, if it occurs in a germ cell, may be passed on to offspring. Some mutations are beneficial, but most are harmful and result in death or disease.

II. Topics for Review

A. Chromosomes
   1. Distribution of chromosomes to offspring
B. Genes and their functions
   1. Dominant and recessive genes
   2. Sex determination
   3. Sex-linked traits

    4. Multifactorial inheritance
    5. Factors influencing gene expression
 C. Mutation

III. Matching Exercises

Matching only within each group, print the answers in the spaces provided.

Group A

    genes   life span   DNA
    chromosomes   carrier   Mendel
    dominant

1. Among the traits influenced by heredity is a person's   _____

2. Independent units of heredity are called   _____

3. A gene that is always expressed if present is described as   _____

4. The person who is credited with the first scientific investigation of heredity was an Austrian monk named   _____

5. A combination of thousands of genes is found in each of the nuclear structures known as   _____

6. The complex molecule that comprises the chemical compound of each chromosome is   _____

7. A person who has a recessive gene that is not expressed is called a(n)   _____

Group B

    DNA   enzymes   hereditary
    RNA   male   female
    mutation   sex-linked   mutagenic
    multifactorial

1. Within the nucleus of each cell there is a genetic chemical that controls all the activities of the cell. This chemical is called   _____

2. Proteins are manufactured in the cytoplasm of the cell under the control of a genetic substance called   _____

3. Traits transmitted by genes are   _____

4. Proteins that promote chemical reactions within cells are   _____

5. If a sperm cell carrying a Y chromosome fertilizes an ovum, the sex of the offspring is   _____

6. Hemophilia is an example of a trait carried on the X chromosome and described as   _____

7. A spontaneous chromosomal change is called a(n) _____

8. A person with two X chromosomes in each cell is a(n) _____

9. Most human traits are determined by two or more pairs of genes acting together in a form of inheritance described as _____

10. A chemical known to produce changes in the genetic material of cells is described as _____

IV. Completion Exercise

Print the word or phrase that most accurately completes each sentence.

1. A spontaneous change in a chromosome is called a(n) _____

2. A gene that is always expressed when it is present in a cell is described as _____

3. In the disorder phenylketonuria, the amino acid phenylalanine cannot be metabolized because a protein is lacking in the cells. This protein functions as a(n) _____

4. Genes are distributed to offspring in the process of cell division that forms the gametes, a process called _____

5. The number of chromosomes in each human germ cell is _____

V. Practical Applications

Study each discussion. Then print the appropriate word or phrase in the space provided.

Group A

These patients were seen in a pediatric clinic.

1. A black child, 4 years of age, was brought to the hospital with a history of swelling and pain in the joints of his hands and feet. Blood studies showed crescent-shaped red blood cells typical of a hereditary disease called sickle cell anemia. This disease is caused by a gene that must be received from both parents in order to appear. Such a gene is described as _____

2. Baby D's face was round; her eyes were close-set and slanted upward at the sides, an appearance typical of a genetic disease called Down's syndrome. This condition is usually caused by an extra strand of DNA in the cells. These strands, which carry the genes, are called _____

3. Baby D's condition was probably caused by an unexplained change in the genetic material of her cells during development. Such a spontaneous change is called a _____

4. Mrs. E began to suspect by the time he was 2 years old that her son had red-green color-blindness. This trait is carried on the X chromosome and is passed from mothers to sons. Any gene that is carried on a sex chromosome is described as _____

5. Mr. and Mrs. S noticed that even though they both had light brown hair their five children had hair color ranging from dark blond to brown. Hair color, as well as many other human traits, is determined by several gene pairs acting together. Traits determined by more than one gene pair are described as _____

6. Mr. and Mrs. C were delighted to give birth to their first child, a girl. The sex of offspring is determined by a pair of chromosomes named X and Y. The genetic makeup of all females is _____

7. Mrs. F and her husband received genetic counseling. They had one normal child and one child that lacked normal skin pigmentation, an albino. They wanted to know whether there was a possibility that a third child would be an albino also. Neither parent was an albino but each had apparently passed on a gene for the trait to one of their children. A person who has a gene that does not appear but can be passed on to offspring is called a _____

21

Biological Terminology

I. Overview

Biological terminology is the special language used worldwide by persons in scientific occupations. Many terms used today originated from Latin or Greek words, but some have come from more recent languages such as French and German. New words are being added constantly as discoveries are made and the need for words to describe them arises. Because scientific knowledge grows in different places at the same time, there may be two or even more terms in use that mean the same thing. Efforts are always being made, however, to standardize the terminology so that people all over the world will "speak the same language."

Not only does biological terminology have universal application, but there are also other advantages to its use. Often someone will say, "Why not use simple plain English?" The fact is that often there is no English word that is as precise as the scientific term. Moreover, one word or perhaps two can do the work of several sentences in descriptive force and accuracy. Biological terminology is a kind of shorthand; workers in scientific occupations should be so familiar with it that it becomes a "second language" with which they feel completely at ease.

Most biological words are made up of two or more parts. The main part is called the root, or the combining form to which the other parts are attached. These other parts include prefixes, which come before the root, and suffixes, which follow the root. If more than one root or combining form plus one or more other parts form the word, it is a compound word. Take the time to divide each medical word into its parts and then look up the meaning of each part, studying each as you go; you will soon add many words to your vocabulary. Then if you practice saying the word, vocalizing each syllable separately, you will feel at ease with biological terminology. Here are some examples:

1. *hypothermia* (hi"po-therm'me-ah): below-normal body temperature, usually due to excessive exposure to cold weather or icy water
   a. prefix (*hypo* = below normal)
   b. root (*therm* = heat)
   c. suffix (*ia* = condition or state of being)

2. *cardiopulmonary* (kar″de-o-pul′mo-nar-e): related to heart and lungs
   a. combining form (*cardio* = heart)
   b. root (*pulmonary* = related to the lungs)
3. *endometrial* (en″do-me-tri′al): pertaining to the lining of the uterus
   a. prefix (*endo* = within)
   b. root (*metr* = related to the uterus)
   c. suffix (*al* = adjective ending)
4. *megakaryocyte* (meg″ah-kar′e-o-site): a giant bone marrow cell that releases blood platelets
   a. prefix (*mega* = large)
   b. combining form (*karyo* = nucleus)
   c. root (*cyte* = cell)

II. Topics for Review

1. common word roots and combining forms, such as

abdomin-, abdomino-	cleid-, cleido-	hyster-, hystero-
aden-, adeno-	cost-	idio-
arthr-, arthro-	cyt-, cyto-	lact-, lacto-
bio-	derm-, ...rma-	leuc- or leuk-, leuko-
bronch-	ente... , entero-	neph-, nephro-
cardi-, cardio-	g...tr-, gastro-	neuro-
cephal-, cephalo-	gynec-, gyneco-	psych-, psycho-
chole-	hem-, hema-, hemato-, hemo-	somat-, somato-
chondr-, chondro-	hist-, histio-	vas-, vaso-

2. common prefixes (... the beginnings of words), such as

a-, an-	inter-	poly-
ab-	intra-	post-
circum-	macro-	semi-
contra-	meg-, mega-, megalo-	sub-
di-	met-, meta-	trans-
ex-	micro-	tri-
infra-	neo-	uni-

3. common sufixes (word endings), such as

-ase	-gen	-oid
-blast	-geny	-phagia, -phagy
-ectasis	-gram	-phil, -philic
-emia	-graph	-pnea
-esthesia	-logy, -ology	-scope
-ferent	-meter	-tropic

4. common adjective endings, such as *-ous* and *-al*
5. common noun endings including *-us* and *-um*

III. Matching Exercises

Matching only within each group, print the answers in the spaces provided.

Group A

prefix	suffix	root
-ous, -al	-logy	a-, an-
compound word	combining form	

1. The foundation of a word is its _____

2. When two or more word foundations are used, the result is a(n) _____

3. The part of a word that precedes its foundations and changes its meaning is a(n) _____

4. A word ending used to change the meaning of the word foundation is a(n) _____

5. Examples of endings that indicate the adjective forms are _____

6. The word root followed by a vowel (to make pronunciation easier) is a(n) _____

7. A suffix that means study of _____

8. To denote absence or deficiency, begin the word with prefixes such as _____

Group B

psych-	abdomin-	trans-
cyt-	hema-	somat-
hist-	aden-	neo-

1. To indicate the belly area, use _____

2. A word root that means gland is _____

3. To show relationship to a cell, use _____

4. A word root for tissue is _____

5. Relationship to mind is shown by _____

6. A word part that means blood is _____

7. A word root that indicates body is _____

8. A prefix that means new is _____

9. A prefix that means through or across is _____

Group C

arthr-   dorso-   dent-
-esthesia   infra-   -blast
meg-   heter-   viscer-

1. A prefix that indicates excessively large is _____

2. A suffix that means an immature cell or early stage is _____

3. A root that means tooth is _____

4. To show relationship to a joint, use _____

5. A root that refers to internal organs is _____

6. A combining form that refers to the back is _____

7. To refer to sensation, use the suffix _____

8. A prefix that means other or different is _____

9. To show that a part is located below, use the prefix _____

Group D

-us, -um   -ous, -al   encephal-
leuko-   -genic   -penia
cardi-   erythr-   ab-
-ia, -ism

1. To indicate producing, add _____

2. A lack of is shown by the suffix _____

3. A combining form that means heart is _____

4. A root that means brain is _____

5. To show that something is red, use the word part _____

6. To indicate that something is white, use _____

7. A prefix that means away from is _____

8. Endings that show the adjective form are _____

9. Noun forms of words may end in _____

10. Endings that mean state of are _____

Group E

Combine appropriate word parts from the list below and print the correct words in the blanks.

hemo-, hemat-, or hemato-	-scope	-logy
-costal	-um	oste- or osteo-
inter-	-lysis	micro-
broncho-	bio-	-cellular
chondri-, or chondro-	peri-	intra-
cyto-, or -cyte		

1. The study of living things is called _____

2. The connective tissue membrane covering a bone is the _____

3. An instrument for studying objects too small to be seen with the eye alone is a(n) _____

4. The scientific study of cells is known as _____

5. An instrument for studying the air passageways of the respiratory system is a(n) _____

6. The space between the ribs is _____

7. A cartilage cell is a(n) _____

8. The study of blood and its constituents is _____

9. The study of organisms too small to be seen with the eye alone is _____

10. The word that means between cells is _____

11. The word that means inside of or within a cell is _____

12. The connective tissue membrane that covers cartilage is _____

13. The dissolution or disintegration of blood cells (especially red blood cells) is called _____

14. Destruction or dissolution of body cells may be called _____

15. A mature bone cell is a(n) _____

IV. Completion Exercise

Print the word or phrase that correctly completes each sentence.

1. A prefix that indicates very small size is _____

2. Words that refer to an instrument for recording end with _____

223

3. The visible record produced by a recording instrument is indicated by a word ending in   _____

4. A prefix that denotes below or under is   _____

5. To show that something is outside or is sent outside, use the prefix   _____

6. To indicate that there are three parts to an organ, begin the word with the prefix   _____

7. A prefix that means across, through, or beyond is   _____

8. To indicate a vessel use the root angio- or   _____

9. The prefix that shows something is within the structure is   _____

10. The noun form of the adjective mucous is   _____

11. Suffixes that indicate the process of eating or swallowing include   _____

12. A combining form that means bladder or sac is   _____

13. The prefix that means away from is   _____

14. A two-letter prefix that means absence or lack of is   _____

15. A suffix that means dilation or expansion of a part is   _____

16. An agent that produces or originates is indicated by the suffix   _____

17. The word root for tissue is   _____

18. Prefixes that mean excessively large include   _____

V. Practical Applications

Study each discussion. Then print the appropriate word or phrase in the space provided.

1. Baby John was brought to the clinic by his observant mother because one of his eyes did not seem normal. The doctor noted that there was unilateral enlargement of the right pupil and that this uniocular condition would require laboratory investigation. The prefix uni- means   _____

2. Mrs. B was admitted for treatment of an injured hand. The admitting intern noted that examination of the metacarpal bones showed possible fractures. The prefix meta- means   _____

3. Ms. C was examined in the outpatient department. It was noted that she had circumoral pallor and that this pallor was circumscribed. The prefix circum- means   _____

4. Baby M was born with an enlarged head due to accumulation of fluid within the skull. This condition is termed hydrocephalus. The word part *hydro-* in this name means _____

5. Mrs. A was admitted for surgery because of bleeding from the uterus. One word root for uterus is metr-; another is _____

6. A first-aid measure everyone should be able to perform involves the heart and the lungs. This type of resuscitation is described by the compound word for heart and lungs, which is _____